A TOUR THROUGH
GRAPH THEORY

TEXTBOOKS in MATHEMATICS

Series Editors: Al Boggess and Ken Rosen

PUBLISHED TITLES CONTINUED

PUBLISHED TITLES CONTINUED

TEXTBOOKS in MATHEMATICS

A TOUR THROUGH
GRAPH THEORY

KARIN R. SAOUB

CRC Press
Taylor & Francis Group
Boca Raton London New York

CRC Press is an imprint of the
Taylor & Francis Group an **informa** business
A CHAPMAN & HALL BOOK

CRC Press
Taylor & Francis Group
6000 Broken Sound Parkway NW, Suite 300
Boca Raton, FL 33487-2742

Version Date: 20171003

International Standard Book Number-13: 978-1-138-19780-0 (Paperback)
International Standard Book Number-13: 978-1-138-07084-4 (Hardback)

Library of Congress Cataloging-in-Publication Data

Names: Saoub, Karin R., author.
Title: A tour through graph theory / Karin R. Saoub.
Description: Boca Raton : Taylor & Francis, 2017. | "A CRC title, part of the
Taylor & Francis imprint, a member of the Taylor & Francis Group, the
academic division of T&F Informa plc." | Includes bibliographical
references and index.
Identifiers: LCCN 2017025808 | ISBN 9781138197800 (paperback)
Subjects: LCSH: Graph theory--Textbooks.
Classification: LCC QA166 .S26 2017 | DDC 511/.5--dc23
LC record available at https://lccn.loc.gov/2017025808

For my children

Contents

Preface

Graph theory has been my passion since senior year of college. I was hooked after just one week of my first course in graph theory. I completely changed my plans post-graduation, choosing to apply to graduate schools and study more mathematics. I found the interplay between rigorous proofs and simple drawings both appealing and a nice break from my more computationally heavy courses. Since becoming a teacher I have found a new appreciation for graph theory, as a concept that can challenge students' notions as to what mathematics is and can be.

You might wonder why I chose to write this book, as there are numerous texts devoted to the study of graph theory. Most books either focus on the theory and the exploration of proof techniques, or contain a chapter or two on the algorithmic aspect of a few topics from graph theory. This book is intended to strike a balance between the two — focus on the accessible problems for college students not majoring in mathematics, while also providing enough material for a semester long course.

The goal for this textbook is to use graph theory as the vehicle for a one-semester liberal arts course focusing on mathematical reasoning. Explanations and logical reasoning for solutions, but no formal mathematical proofs, are provided. There are discussions of both historical problems and modern questions, with each chapter ending in a section detailing some more in-depth problems. The final chapter also provides more rigorous graph theory and additional topics that I personally find interesting but for which there is not enough material to warrant an entire chapter. Each chapter will include problems to test understanding of the material and can be used for homework or quiz problems. In addition, project ideas and items requiring research are included at the end of each exercise section. Selected answers are available on page 289.

Advice for Students

Reading a mathematics textbook takes skill and more effort than reading your favorite novel at the beach. Professors often complain that their students are not getting enough out of the readings they assign, but fail to realize that most students have not been taught how to read mathematics. My advice can be boiled down to one sentence: anytime you read mathematics, have paper and pencil next to your book. You should expect to work through examples, draw graphs, and play around with the concepts. We learn by doing, not passively reading or watching someone perform mathematics. This book contains examples often posed in the form of a question. You should attempt to find the solution before reading the one provided. In addition, some definitions and concepts can get technical (as happens in mathematics) and the best way to truly understand these is through working examples. At times, details of an example, especially if it is the second or third of a type, will be left for the reader or will appear in the exercises.

Advice for Instructors

This book is split into three main areas. The first focuses on topics related to tours within a graph (Eulerian circuits, Hamiltonian cycles, and shortest paths), the second to finding structure within a graph (spanning trees, maximum matchings, and optimal coloring), and the third is a collection of additional topics that can be used to delve deeper into topics covered in previous chapters or topics that do no warrant an entire chapter. Definitions will appear as needed, but as with most mathematics books, later chapters build upon previously introduced concepts. The chart below outlines how the sections are related.

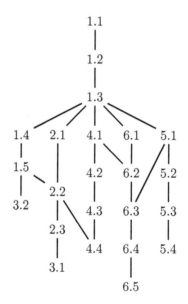

The Additional Topics sections in Chapter 7 may relate to multiple earlier sections and will indicate individually which prior material is necessary. References to these sections will occasionally appear throughout the first six chapters of the book.

Thanks

I owe a huge debt to the many people who made this book possible.

To my colleagues Adam Childers, Chris Lee, Roland Minton, Maggie Rahmoeller, Hannah Robbins, and David Taylor for their unwavering support. In particular, David Taylor who provided a great sounding board for my ideas, gave advice on formatting, and found the typos I could not.

To my sister-in-law, Leena Saoub Saunders, for her expertise in designing the cover of this book.

To my parents, David and Pamela Steece, who always supported my dreams and taught me the benefit of hard work.

To my husband Samer and children, Layla and Rami, who bring joy to my life and endured my excitement and exasperation while writing this book.

To my CRC Press team, namely editor Bob Ross, editorial assistant Jose Soto, copy editor Michele Dimont, and designer Kevin Craig, for their guidance through the long process of writing a book.

Finally, a very special thanks goes to Ann Trenk, my graph theory professor at Wellesley College. If not for her, I would not have discovered my life's passion and you would not be reading this book.

Dr. Karin R. Saoub
MCSP Department
Roanoke College
Salem, Virginia 24153
saoub@roanoke.edu

Chapter 1

Eulerian Tours

1.1 Königsberg Bridge Problem

You arrive in a new city and hear of an intriguing puzzle captivating the population: can you leave your home, travel across each of the bridges in the city exactly once and then return home? Upon one look at the map, you claim "Of course it can't be done!" You describe the requirements needed for such a walk to take place and note this city fails those requirements. Easy enough!

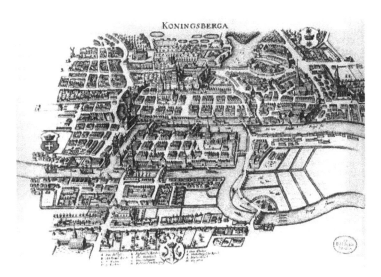

The puzzle above is described by some as the birth of graph theory. In 1736, Leonhard Euler, one of the greatest mathematicians of all time, published a short paper on the bridges of Königsberg, a city in Eastern Europe (see the map above). Euler translated the problem into one of the "geometry of location" (*geometris situs*) and determined that only certain configurations would allow a solution to be possible. His publication set in motion an entirely new branch of mathematics, one that has profound impact in modern mathematics, computer science, management science, counterterrorism, ... and the list continues. This book will explore some of these topics, but we begin with some basic terminology and mathematical modeling.

Without worrying about the technicalities, see if you can find a solution to the Königsberg Bridge Problem! To aid in your analysis, note that Königsberg contains seven brides and four distinct landmasses.

1.2   Introduction to Graph Models

Definition 1.1 A ***graph*** G consists of two sets: $V(G)$, called the vertex set, and $E(G)$, called the edge set. An ***edge***, denoted xy, is an unordered pair of vertices. We will often use G or $G = (V, E)$ as short-hand.

In a general graph xy and yx are treated equally, though it is customary to write them in alphabetical order. In Chapter 2 we will study directed graphs where the order in which an edge is written provides additional meaning.

Example 1.1 Let G be a graph where $V(G) = \{a, b, c, d, e\}$ and $E(G) = \{ab, cd, cd, bb, ad, bc\}$. Although G is defined by these two sets, we generally use a visualization of the graph where a dot represents a vertex and an edge is a line connecting the two dots (vertices). A drawing of G is given below.

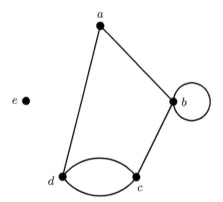

Note that two lines were drawn between vertices c and d as the edge cd is listed twice in the edge set. In addition, a circle was drawn at b to indicate an edge (bb) that starts and ends at the same vertex.

It should be noted that the drawing of a graph can take many different forms while still representing the same graph. The only requirement is to faithfully record the information from the vertex set and edge set. We often draw graphs with the vertices in a circular pattern (as shown in the example above), though in some instances other configurations better display the desired information. The best configuration is the one that reduces complexity or best illustrates the relationships arising from the vertex set and edge set.

Example 1.2 Consider the graph G from Example 1.1. Below are two different drawings of G.

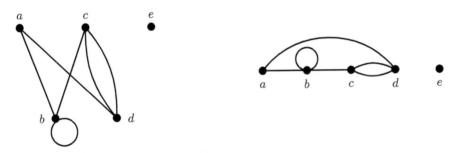

To verify that these drawings represent the same graph from Example 1.1, we should check the relationships arising from the vertex set and edge set. For example, there are two edges between vertices c and d, a loop at b, and no edges at e. You should verify the remaining edges.

To discuss the properties of graphs in mathematical models, we need the proper terminology. The graph given in the examples above works as a good reference for this initial terminology. The formal definitions are given below, followed by the appropriate references to the graph in Example 1.1 (or Example 1.2).

Definition 1.2 Let G be a graph.

- If xy is an edge, then x and y are the **endpoints** for that edge. We say x is **incident to** edge e if x is an endpoint of e.

- If two vertices are incident to the same edge, we say the vertices are **adjacent**. Similarly, if two edges share an endpoint, we say they are adjacent. If two vertices are adjacent, we say they are **neighbors** and the set of all neighbors of a vertex x is denoted $N(x)$.

  - ab and ad are adjacent edges since they share the endpoint a
  - a and b are adjacent vertices since ab is an edge of G
  - $N(d) = \{a, c\}$ and $N(b) = \{a, b, c\}$

- If a vertex is not incident to any edge, we call it an **isolated vertex**.

  - e is an isolated vertex

- If both endpoints of an edge are the same vertex, then we say the edge is a **loop**.

- *bb* is a loop

- If there is more than one edge with the same endpoints, we call these *multi-edges*.

  - *cd* is a multi-edge

- If a graph has no multi-edges or loops, we call it *simple*.

- The *degree* of a vertex is the number of edges incident to that vertex, with a loop adding two to the degree. Denote the degree of vertex v as $\deg(v)$. If the degree is even, the vertex is called *even*; if the degree is odd, then the vertex is *odd*.

  - $\deg(a) = 2$, $\deg(b) = 4$, $\deg(c) = 3$, $\deg(d) = 3$, $\deg(e) = 0$

Now that we have some basic graph terms, let's look back at the Königsberg Bridge Problem. Euler brilliantly reduced the map of Königsberg to a simpler version where only the relationships between landmasses was of importance. In graph form, the vertices represent the landmasses in the city and the bridges are the edges. This is what we refer to as *modeling* — taking a complex real-world problem and representing it mathematically.

Modeling is used in many branches of mathematics, but often the phenomenon in question is being translated into an equation or experiment; here we are representing a puzzle as a graph and using properties of graphs to find a solution. To the right is the drawing from Euler's original paper, and below it a graph model of Königsberg. You should see how much simpler the graph is compared to either the original map or Euler's drawing of the city. This is one benefit of using a graph — only the relevant information is displayed, which in this case consists solely of bridges connecting landmasses. The vertices represent an island, a south bank, a north bank, and a peninsula (a, b, c, and d, respectively). The answer to the initial question becomes more evident when using the graph: can you leave your home (vertex), travel across each of the bridges (edges) in the city (graph) exactly once and then return home? If you have not done so already, try it again on the graph. Do you have an answer?

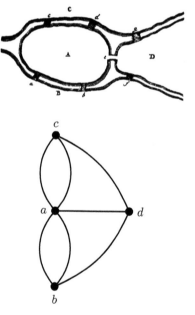

Before we get too much further into answering the Königsberg Bridge

Problem and its subsequent generalization, let's go through a few more modeling problems.

Example 1.3 Professor Minton needs to create a seating chart for his overly chatty class. He wants students seated away from their friends whenever possible. He knows that Adam is friends with Betty, Carlos, Dave and Frank; Betty is friends with Adam, Carlos, Dave and Emily; Dave is friends with Adam, Betty, Carlos and Emily; and Frank is friends with Adam and Carlos. Draw a graph that depicts these friendships.

Solution: Each person is represented by a vertex and an edge denotes a friendship.

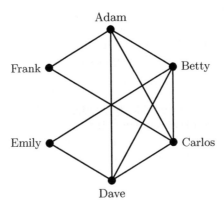

Note that since friendships are symmetric (e.g., Adam is friends with Betty and so Betty is friends with Adam) we only draw one edge between the vertices. In Chapter 2 we will explore asymmetric relationships where directionality becomes important.

As the example above demonstrates, problems that can be modeled by a graph need to consist of distinct objects (such as people) and a relationship between them (such as a friendship). The proper model will allow the graph structure, or properties of the graph, to answer the question being asked. The example below gives a different approach to graph modeling.

Example 1.4 Three student organizations (Student Government, Math Club, and the Equestrian Club) are holding meetings on Thursday afternoon. The only available rooms are 105, 201, 271, and 372. Based on membership and room size, the Student Government can only use 201 or 372, Equestrian Club can use 105 or 372, and Math Club can use any of the four rooms. Draw a graph that depicts these restrictions.

Solution: Each organization and room is represented by a vertex, and an edge denotes when an organization is able to use a room.

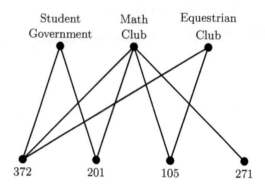

Note that edges do not occur between two organizations or between two rooms, as these would be nonsensical in the context of the problem. The graph above is an example of a *bipartite graph*. This type of graph is often used to model interactions between two distinct types of groups and will appear later in Chapters 5 and 6.

Hopefully, you now have a clearer understanding of how modeling works. We often need to think about what information is relevant, what question we are trying to answer, and what type of visual display to employ. See the exercises at the end of the chapter for more practice with graph modeling.

1.3   Touring a Graph

A part of the modeling process is determining what the answer we are searching for looks like in graph form. What type of structure or operation are we looking for? In our original search for a solution to the Königsberg Bridge Problem, we discuss traveling through the city. What would traveling through a graph mean? What types of restrictions might we place on such travel? Below are additional definitions needed to answer these questions, after which we will finally describe how to use a graph model to solve the Königsberg Bridge Problem. As with the previous definitions, you are encouraged to refer back to the graph in Example 1.1 for examples of these terms.

Definition 1.3 Let G be a graph.

- A *walk* is a sequence of vertices so that there is an edge between consecutive vertices. A walk can repeat vertices and edges.

- A *trail* is a walk with no repeated edges. A trail can repeat vertices but not edges.

- A **path** is a trail with no repeated vertex (or edges). A path on n vertices is denoted P_n.

- A **closed walk** is a walk that starts and ends at the same vertex.

- A **circuit** is a closed trail; that is, a trail that starts and ends at the same vertex with no repeated edges though vertices may be repeated.

- A **cycle** is a closed path; that is, a path that starts and ends at the same vertex. Thus cycles cannot repeat edges or vertices. Note: we do not consider the starting and ending vertex as being repeated since each vertex is entered and exited exactly once. A cycle on n vertices is denoted C_n.

The **length** of any of these tours is defined in terms of the number of edges. For example, P_n has length $n - 1$ and C_n has length n.

Technically, since a path is a more restrictive version of a trail and a trail is a more restrictive form of a walk, any path can also be viewed as a trail and as a walk. However, a walk might not be a trail or a path (for example, if it repeats vertices or edges). Similarly, a cycle is a circuit and a closed walk. Unless otherwise noted, when we use any of the terms from Definition 1.3 we are referring to the most restrictive case possible; for example, if we ask for a walk in a graph then we want a walk that is not also a trail or a path.

In practice, it is often necessary to label the edges of a tour of a graph in the sequential order in which they are traveled. This is especially important when the graph is not simple.

Example 1.5 Given the graph below, find a trail (that is not a path) from a to c, a path from a to c, a circuit (that is not a cycle) starting at b, and a cycle starting at b.

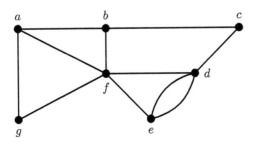

Solution:

Trail from *a* to *c*

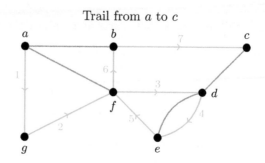

Path from *a* to *c*

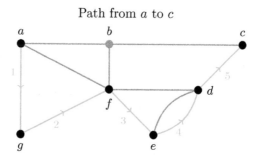

Circuit starting at *b*

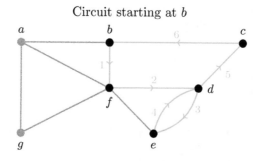

Cycle starting at *b*

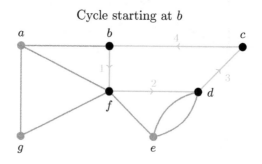

Note that the trail from a to c is not a path since the vertex f is repeated and the circuit starting at b is not a cycle since the vertex d is repeated. Moreover, the examples given above are not the only solutions but rather an option among many possible solutions.

Earlier we defined what it meant for two vertices to be *adjacent*, namely x and y are adjacent if xy is an edge in the graph. Notice that in common language we may have wanted to say that x and y are connected since there is a line that connects these dots on the page. However, in graph theory the term *connected* refers to a different, though related, concept.

Definition 1.4 Let G be a graph. Two vertices x and y are **connected** if there exists a path from x to y in G. The graph G is **connected** if every pair of distinct vertices is connected.

This definition may seem overly technical when visually it is often easy to determine if a graph is connected. The concept of connectedness is surprisingly important in applications. The example below illustrates the importance of connectedness for the Königsberg Bridge Problem.

Example 1.6 Island City has two islands, a peninsula, and left and right banks, as shown on the left below. Modeling the relationship between the landmasses and bridges of Island City gives us the graph below on the right.

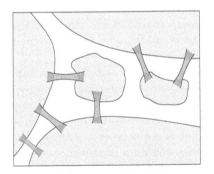 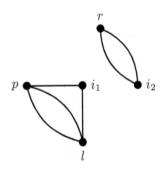

If the same puzzle were posed to the citizens of Island City as the one for Königsberg, it would be impossible to travel every bridge and then return home. In particular, there is no way to even travel from one side of the river to the other! Thus, if we even have a hope of finding a solution, the resulting graph must, at a *minimum*, be connected.

The Königsberg Bridge Problem is not just looking for a circuit, but rather a special type that hits every edge. In honor of Leonhard Euler, these special circuits are named for him. A similar term is used when allowing differing starting and ending location of the tour.

Definition 1.5 Let G be a graph. An ***Eulerian circuit*** (or ***trail***) is a circuit (or trail) that contains every edge and every vertex of G. If G contains an Eulerian circuit it is called ***Eulerian*** and if G contains an Eulerian trail but not an Eulerian circuit it is called ***semi-Eulerian***.

Part of Euler's brilliance was not only his ability to quickly solve a puzzle, such as the Königsberg Bridge Problem, but also the foresight to expand on that puzzle. What *makes* a graph Eulerian? Under what conditions will a city have the proper tour? In his original paper, Euler laid out the conditions for such a solution, though as was typical of the time, he only proved a portion of the statement (see [2] or [14]).

Theorem 1.6 A graph G is Eulerian if and only if

(i) G is connected and

(ii) every vertex has even degree.

A graph G is semi-Eulerian if and only if

(i) G is connected and

(ii) exactly two vertices have odd degree.

The theorem above is of a special type in mathematics. It is written as an "*if and only if*" statement, which indicates that the conditions laid out are both necessary and sufficient. A *necessary condition* is a property that must be achieved in order for a solution to be possible and a *sufficient condition* is a property that guarantees the existence of a solution.

For a more familiar example, consider renting and driving a car. If you want to rent a car, a necessary condition would be having a driver's license; but this condition may not be sufficient since some companies will only rent a car to a person of at least 25 years of age. In contrast, having a driver's license is sufficient to be able to drive a car, but is not necessary since you can drive a car with a learner's permit as long as a guardian is present.

Mathematicians often search for a property (or collection of properties) that is both necessary and sufficient (such as a number is even if and only if it is divisible by 2). The theorem above gives both necessary and sufficient conditions for a graph to be Eulerian or semi-Eulerian. It should be clear why connectedness must be achieved if every vertex is to be reached in a single tour. Can you explain the degree condition? When traveling through a graph, we need to pair each entry edge with an exit edge. If a vertex is odd, then there is no pairing available and we would eventually get stuck at that vertex. However, if the starting and ending locations are different, then exactly two vertices must be odd since the first edge out of the starting vertex does not need to be paired with a return edge and the last edge to the ending vertex does not need to be paired with an exit edge.

The proof that the conditions above were indeed both necessary and sufficient was not published until 1873. The work was completed by German mathematician Carl Hierholzer, who unfortunately died too young to see his work in print. His contribution was recognized through the naming of one procedure for finding Eulerian circuits that is discussed in the next section.

Example 1.7 Look back at each of the graphs appearing so far in this chapter. Which ones are Eulerian? semi-Eulerian? neither?

Solution:

- The graph representing Königsberg is neither Eulerian nor semi-Eulerian since all four vertices are odd.

- The graph in Example 1.1 is neither Eulerian nor semi-Eulerian since it is not connected.

- The graph in Example 1.3 is Eulerian since it is connected and all the vertices have degree 2 or 4.

- The graph in Example 1.4 is semi-Eulerian since it is connected and exactly two vertices are odd (namely, 372 and 271).

- Even though the graph in Example 1.5 is connected, it is neither Eulerian nor semi-Eulerian since it has more than two odd vertices (namely, $a, b, e,$ and f).

Note that the solution discussed in the first bullet above definitively answers the Königsberg Bridge Problem — there is no way to leave your home, travel across every bridge in the city exactly once, and return home!

In addition to the result above, Euler also determined some basic properties of graphs in his seminal paper. The most important of these is stated below.

Theorem 1.7 (Handshaking Lemma) Let $G = (V, E)$ be a graph and $|E|$ denote the number of edges in G. Then the sum of the degrees of the vertices equals twice the number of edges; that is if $V = \{v_1, v_2, \ldots, v_n\}$, then

$$\deg(v_1) + \deg(v_2) + \cdots + \deg(v_n) = 2|E|$$

This is often referred to as the Handshaking Lemma, as a quick proof of the statement is best described by a graph model of handshakes. Define a graph where each vertex represents a person and an edge represents a handshake that has occurred between the people represented by the endpoints of the edge. Then the total number of handshakes is equal to the total number of edges in the graph and the number of times a person shook hands is equal to the degree of his or her representative vertex. When totaling the degrees of all the vertices, each handshake will be counted twice (one for each person involved) and so the sum of the degrees equals twice the number of edges.

If the sum total of a collection of integers is even, how many odd integers could appear in the sum? If not immediately clear, try a few examples. If we add an odd number of odd integers, will the sum be even? No! In graph theoretic terms, we get the following corollary to the Handshaking Lemma.

Corollary 1.8 There must be an even number of odd vertices in any graph G.

Why is this true? Since the sum of the degrees in any graph is twice the number of edges, we know that the sum of the degrees must be even. Based on our discussion above, this implies the number of odd vertices must also be even.

1.4   Eulerian Circuit Algorithms

The previous section was mainly concerned with the *existence* question — under what conditions will a graph have an Eulerian circuit? This was definitively answered with two conditions that are easy to check. This section focuses on the *construction* question: how do we find an Eulerian circuit once we know one exists?

There are numerous methods for finding an Eulerian circuit (or trail), though we will focus on only two of these. Each of these is written in the form of an *algorithm*.

Definition 1.9 [36] An **algorithm** is a procedure for solving a mathematical problem in a finite number of steps that frequently involve repetition of an operation.

For our purposes, algorithms will be described in terms of the input, steps to perform, and output, so it is clear how to apply the algorithm in various scenarios. For the pseudo-code or more technical forms of the algorithms in this book, the reader is encouraged to explore [4] or [24].

Fleury's Algorithm

The first method for finding an Eulerian circuit that we discuss is Fleury's Algorithm. Although Fleury's solution was not the first in print, it is one of the easiest to walk through (no pun intended) [16]. As with all future algorithms presented in this book, an example will immediately follow the description of the algorithm and further examples are available in the exercises. Note that Fleury's Algorithm will produce either an Eulerian circuit or an Eulerian trail depending on which solution is possible.

Fleury's Algorithm

<u>Input</u>: Connected graph G where zero or two vertices are odd.

<u>Steps</u>:

1. Choose a starting vertex, call it v. If G has no odd vertices, then any vertex can be the starting point. If G has exactly two odd vertices, then v must be one of the odd vertices.

2. Choose an edge incident to v that is unlabeled and label it with the number in which it was chosen, ensuring that the graph consisting of unlabeled edges remains connected.

3. Travel along the edge to its other endpoint.

4. Repeat steps 2 and 3 until all edges have been labeled.

<u>Output</u>: Labeled Eulerian circuit or trail.

The intention behind Fleury's Algorithm is that you are prevented from getting stuck at a vertex with no edges left to travel. In practice, it may be helpful to use two copies of the graph — one to keep track of the route and the other where labeled edges are removed. This second copy makes it easier to see which edges are unavailable to be chosen. In the example below, the vertex under consideration during a step of the algorithm will be highlighted and edges will be labeled in the order in which they are chosen.

Example 1.8 Input: A connected graph (shown below) where every vertex has even degree. We are looking for an Eulerian circuit.

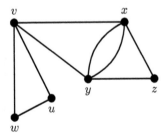

Step 1: Since no starting vertex is explicitly stated, we choose vertex v to be the starting vertex.

Step 2: We can choose any edge incident to v. Here we chose vx. The labeled graph is on the left and the unlabeled portions are shown on the right with edges removed that have already been chosen.

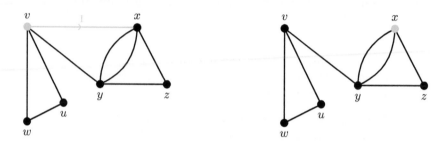

Step 3: Looking at the graph to the right, we can choose any edge out of x. Here we chose xy. The labeled and unlabeled graphs have been updated below.

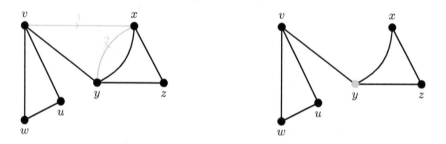

Step 4: At this point we cannot choose yv, as its removal would disconnect the unlabeled graph shown above on the right. However, yx and yz are both valid choices. Here we chose yx.

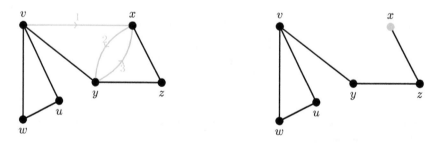

Step 5: There is only one available edge xz.

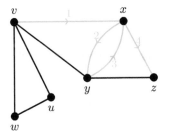 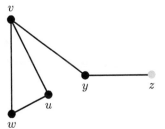

Step 6: There is only one available edge *zy*.

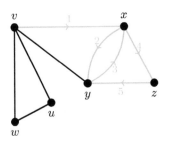 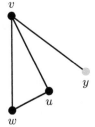

Step 7: There is only one available edge *yv*.

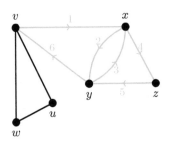 

Step 8: Both *vw* and *vu* are valid choices for the next edge. Here we chose *vw*.

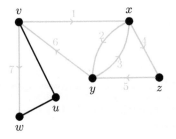 

Step 9: There is only one available edge wu.

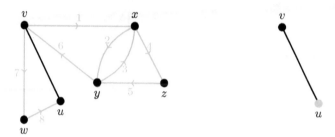

Step 10: There is only one available edge uv.

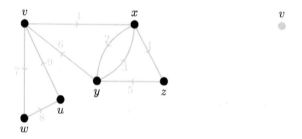

Output: The graph above on the left has an Eulerian circuit labeled, starting and ending at vertex v.

Hierholzer's Algorithm

As mentioned above, Hierholzer's Algorithm is named for the German mathematician whose paper inspired the procedure described below. This efficient algorithm begins by finding an arbitrary circuit originating from the starting vertex. If this circuit contains all the edges of the graph, then an Eulerian circuit has been found. If not, then we join another circuit to the existing one.

Hierholzer's Algorithm

Input: Connected graph G where all vertices are even.

Steps:

1. Choose a starting vertex, call it v. Find a circuit C originating at v.

2. If any vertex x on C has edges not appearing in C, find a circuit C' originating at x that uses two of these edges.

3. Combine C and C' into a single circuit C^*.

4. Repeat steps 2 and 3 until all edges of G are used.

Output: Labeled Eulerian circuit.

Note that Hierholzer's Algorithm requires the graph to be Eulerian, whereas Fleury's Algorithm allows for the graph to be Eulerian or semi-Eulerian. In the implementation of Hierholzer's Algorithm shown below, a new circuit will be highlighted in blue with other edges in gray. As with Fleury's Algorithm, the edges will be labeled in the order in which they are traveled.

Example 1.9 Input: A connected graph where every vertex has even degree. We are looking for an Eulerian circuit.

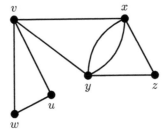

Step 1: Since no starting vertex is explicitly stated, we choose v and find a circuit originating at v. One such option is highlighted below.

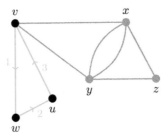

Step 2: As $\deg(v) = 4$ and two edges remain for v (shown in gray above), a second circuit starting at v is needed. One option is shown below.

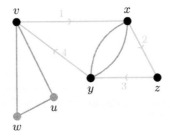

Step 3: Combine the two circuits from Step 1 and Step 2. There are multiple ways to combine two circuits, but it is customary to travel the first circuit created and then travel the second.

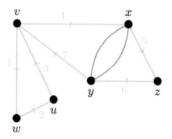

Step 4: As $\deg(x) = 4$ and two edges remain for x (shown in gray above), a circuit starting at x is needed. It is shown below.

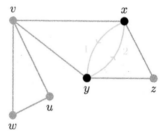

Step 5: Combine the two circuits from Step 3 and Step 4.

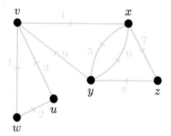

Output: The graph above gives a labeled Eulerian circuit originating at v.

There are advantages and disadvantages for these two methods; in particular, both Fleury's and Hierholzer's Algorithms will find an Eulerian circuit when one exists, whereas only Fleury's can be used to find an Eulerian trail (see the exercises for a modification of Hierholzer's that will find an Eulerian trail). Try a few more examples and make additional comparisons. In practice, either algorithm is a good choice for finding an Eulerian circuit — pick the method that works best for you.

Barring these differences, the two Eulerian algorithms are fairly *efficient*.

Algorithm efficiency will be discussed in more detail in the next chapter (as well as in Section 7.1), but for now think of algorithm efficiency as measuring how much time is needed for an algorithm to find a solution. Efficiency takes into account the number of calculations needed to run the algorithm as the size of the problem grows. An algorithm is considered efficient if the run time grows at roughly the same speed as the size of the graph. For example, applying Fleury's Algorithm to a graph with 25 vertices is not that much more difficult than a graph on 10 vertices. An inefficient algorithm is one in which the run time grows much faster than the size of the graph.

At this point, we have answered both the existence and construction questions. But what happens when a graph does not have an Eulerian circuit? Do we give up and move on? Obviously, the answer is no. The next section will introduce a third question: *optimization.*

1.5   Eulerization

Look back at the graph from Example 1.5. We have already determined that there is no Eulerian circuit or Eulerian trail due to four odd vertices in the graph. If this graph modeled a city (with quite a lot of bridges!) then we could not make a tour like the one of interest in Königsberg.

This section looks at how to adjust a graph to ensure an Eulerian circuit or trail can be found. There are two ways a graph will fail to be Eulerian (or semi-Eulerian): the graph is disconnected or the graph contains too many odd vertices. The processes described below focus on the degree condition.

Definition 1.10 Given a connected graph $G = (V, E)$, an **Eulerization** of G is the graph $G' = (V, E')$ so that

(i) G' is obtained by duplicating edges of G, and

(ii) every vertex of G' is even.

A **semi-Eulerization** of G results in a graph G' so that

(i) G' is obtained by duplicating edges of G, and

(ii) exactly two vertices of G' are odd.

Once the Eulerization of a graph G has been completed, the new graph G' satisfies the conditions necessary for an Euler circuit to exist, allowing the algorithms from the previous section to be applied. Viewing this in terms of the Königsberg Bridge Problem, duplicating an edge would be akin to walking the same bridge twice. Although this solution would be outside the original parameters (walking each bridge exactly once), it allows for an approximate solution using the fewest number of duplications. This is often referred to as finding an **optimal exhaustive tour** of a graph.

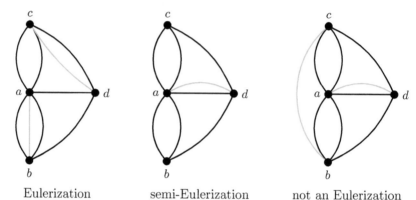

Eulerization   semi-Eulerization   not an Eulerization

Note that when creating the new graph, edges must be duplicated, not added. In Königsberg, creating a new edge in the graph model corresponds to building a new bridge within the city. While this certainly can happen within a city over time (and in fact the bridges in Königsberg have changed considerably over the past one hundred fifty years, see [31]), it is unrealistic from the standpoint of a person touring the bridges of a city during a specific moment in time. Hence, we only allow duplications of edges.

Since the Eulerization (or semi-Eulerization) only allows for edge duplication, we cannot consider disconnected graphs. To make such a graph connected, we would need to add new edges between the individual pieces. Again, in terms of the original problem being modeled, we would be creating bridges that do not already exist (see Example 1.6).

We are now investigating the *optimization* question: how can we Eulerize (or semi-Eulerize) a graph using the fewest number of edge duplications? We are attempting to find an optimal tour of the graph, that is minimize the total length of the circuit (or trail).

Unlike the processes for determining if a graph is Eulerian and then finding an Eulerian circuit, Eulerizing a graph can be quite complicated and the formal algorithms are beyond the scope of this book. Instead, we will discuss a process for Eulerizing a graph and provide examples that display the complexity involved. Portions of this will be revisited in Chapter 3.

Eulerization Method

1. Identify the odd vertices of the graph.

2. Pair up the odd vertices, trying to pair as many adjacent vertices as possible while also avoiding pairing vertices far away from each other.

3. Duplicate the edges along an optimal path from one vertex to its pair.

In the process of determining which edges to duplicate along optimal paths, never repeat an edge more than once. If an edge is crossed three times, removing two of the duplications will not change the parity of the endpoint of the edge; that is, a vertex will remain odd or remain even when subtracting two from the degree.

Example 1.10 The citizens of the small island town of Sunset Island want to hire a night patrol during the busy summer tourist season. A map of the town is shown below. Model the town as a graph and find an optimal Eulerization of the graph.

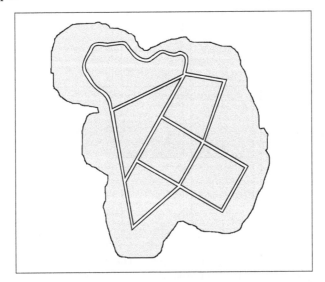

Solution: Below is the graph modeling the town of Sunset Island where vertices represent intersections and edges represent street blocks. Note that there are four odd vertices (namely a, i, j and k). Since these four odd vertices can be split into two pairs of adjacent vertices (a and i, j and k), we can Eulerize the graph using only two edge duplications. An optimal Eulerization is shown in blue.

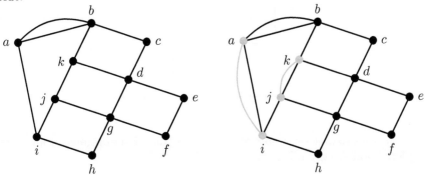

The example above, though small and with just a few odd vertices, demonstrates the general procedure of Eulerizing a graph. The example below provides more complexity and insight into how to handle multiple odd vertices. Note that these examples all focus on Eulerizing a graph; semi-Eulerization will be addressed in the exercises.

Example 1.11 You have been hired to put up fliers along each block of a small portion of your hometown. You need to travel along each street once but cannot put up fliers in the park. Model the situation as a graph and find an optimal Eulerization.

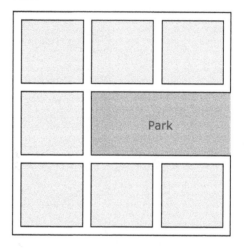

Solution: As above, we begin by modeling the map as a graph and identifying the odd vertices, shown in blue below.

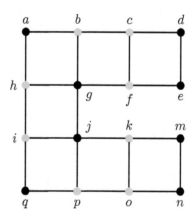

We want to pair the 8 odd vertices into 4 pairs. Four options for an optimal Eulerization are given below.

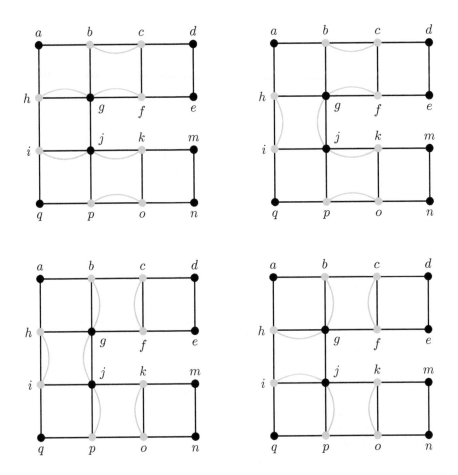

Note that each of these Eulerizations uses exactly 6 duplicated edges. Unlike Example 1.10, we cannot split all the odd vertices into adjacent pairs meaning we will need more than 4 edges in an optimal Eulerization. In addition, we cannot use 5 edges in any Eulerization since if we use exactly 2 adjacent odd pairs (and thus needing 2 edge duplications) we are left with 2 pairs each of which is of distance 2 apart and thus requiring another 4 edge duplications; if we use 3 adjacent pairs, we are left with 1 pair with vertices of distance 3 apart. Both of these require a total of 6 duplicated edges.

Both of the examples above treated the edges equally. For example, in the graph from Example 1.11 we could have duplicated either edges fg, gh, jk and ji or edges fg, gj, jk and hi to make the vertices f, h, i and k even. There is no distinction between the edges and so either choice is optimal. What if the edges are not equivalent? Can you think of a reason why one duplication might be preferable over another?

Chinese Postman Problem

Consider the graph model of Sunset Island shown in Example 1.10 above. Portions of the town have a very regular grid structure and so traveling down one block versus another is inconsequential (think of Manhattan or Phoenix). However, for cities with more of an evolutionary development (such as Boston or Providence) or in rural towns where roads curve and blocks have different lengths. Traveling down a stretch of road twice could look remarkably different from one choice to the next. How then would you model these differences? One solution might be to draw the edges in the graph with a length proportional to the length of the road they model, but how do you code this information into an algorithm? A better method would be to assign numbers to each edge that correspond to the length of the road. These types of graphs are called *weighted graphs*.

Definition 1.11 A *weighted graph* $G = (V, E, w)$ is a graph where each of the edges has a real number associated with it. This number is referred to as the *weight* and denoted $w(xy)$ for an edge xy.

Note that a weighted graph can also refer to a graph in which each of the vertices is assigned a weight, and denoted $w(v)$ for a vertex v. In the next few chapters, we will focus on graphs in which the edges are weighted; the vertex version will be addressed in Chapter 6. Also, the weight associated with an edge can represent more than just length. For example, we may be interested in time or cost as opposed to distance. Choose the appropriate measure based upon the scenario in question.

The weighted version of an Eulerization problem is called the ***Chinese Postman Problem***. The name originates not from anything particular about postmen in China, but rather from the mathematician who first proposed the problem — the Chinese mathematician Mei-Ku Kwan in 1962. This problem first appeared more than two centuries after Euler's original paper! And its full solution was published about a decade later (the impressive, though complicated, algorithm can be found in [13]). The main idea is that a Postman delivering mail in a rural neighborhood should repeat the shortest stretches of road (provided any duplications are necessary).

Example 1.12 Even though the patrolman on Sunset Island enjoys his evening strolls, he would like to complete a circuit through the town in as little time as possible. The weights on the graph below represent the average time it takes the patrolman to travel that stretch of road. Find an optimal Eulerization taking into account these weights.

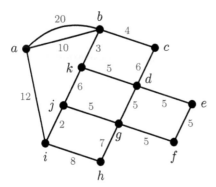

Solution: The Eulerization from Example 1.10 is shown below on the left. Note that the two duplicated edges total 18. A better Eulerization is on the right which duplicates three edges for a total of 15.

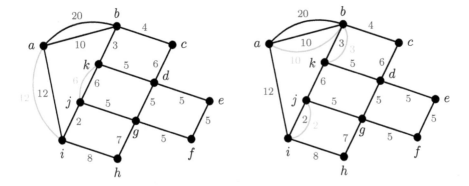

Be careful when duplicating edges in a graph with multi-edges, indicating which edge has been duplicated. The graph on the right duplicated one of the edges between a and b, and including the edge weight clarifies which option was used.

On a general graph, solving the Chinese Postman Problem can be quite challenging. However, most small examples can be solved by inspection since there are relatively few choices for duplicating edges. Would you duplicate 3 edges of weight 1 or one edge of weight 10? The choice should be obvious. In addition, if the weight of an edge represents distance, then we can rely on the real world properties of distance. For example, the shortest path between two points is a straight line and no one side of a triangle is longer than the sum of the other two (this is called the "triangle inequality"). These two properties would eliminate many options when the weight of an edge models distance along a road. The more difficult (and hence more interesting) problems occur when the weight represents something other than distance. Such an example is shown below.

Example 1.13 Find an optimal Eulerization for the graph below where the weights given are in terms of cost.

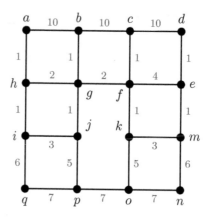

Solution: The optimal Eulerization uses 7 edge duplications with an added weight total of 13 and is shown in blue below.

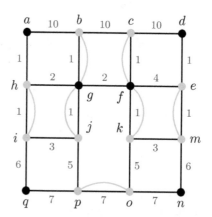

The Eulerization for the unweighted graph would only use 5 edge duplications (try it!). On a weighted graph, we attempt to pair vertices along paths using edges of weight 1 as much as possible. We are able to do this for all vertices except o and p, and we duplicate edge op of weight 7. Any attempt to avoid using this edge would still require both o and p to be paired with another vertex (since they are both odd vertices) and would require the use of edges ko and jp, both of which have weight 5. There is no way to pair the remaining odd vertices while maintaining a total increase in weight of 13.

The choice of which edges to duplicate when working with a weighted graph relies in part on shortest paths between two vertices. The difficulty is in choosing which vertices to pair. This will be revisited in Chapter 3 when an algorithm for finding a shortest path is introduced.

1.6 Exercises

1.1 Let G be a graph with vertex set $V(G) = \{a, b, c, d, e\}$ and edge set $E(G) = \{ab, ae, bc, cd, de, ea, eb\}$.
   (a) Draw G.
   (b) Is G connected?
   (c) Is G simple?
   (d) List the degrees of every vertex.
   (e) Find all edges incident to b.
   (f) List all the neighbors of a.
   (g) Find a walk, trail, and path in G, each of which has length 3.
   (h) Find a closed walk, circuit, and cycle in G, each of which starts at e.
   (i) Is G Eulerian, semi-Eulerian, or neither? Explain your answer.

1.2 Which of the following scenarios could be modeled using an Eulerian circuit? Explain your answer.
   (a) A photographer wishes to visit each of the seven bridges in a city, take photos, then return to his hotel.
   (b) Salem Public Works must repave all the streets in the downtown area.
   (c) Frank's Flowers needs to deliver bouquets to 6 customers throughout the city, starting and ending at the flower shop.
   (d) Richmond Water Authority must read all the water meters throughout the town. One worker is tasked with this job.
   (e) Sam works in sales for a Fortune 500 company. He spends each day visiting his clients around southwest Virginia and must plan his route to avoid backtracking as much as possible.

1.3 Use Theorem 1.6 to explain why a graph cannot be both Eulerian and semi-Eulerian.

1.4 For each of the graphs below (i) find the degree of each vertex, and (ii) use your results from (i) to determine if the graph is Eulerian, semi-Eulerian, or neither. Explain your answer.

(a)

(b)

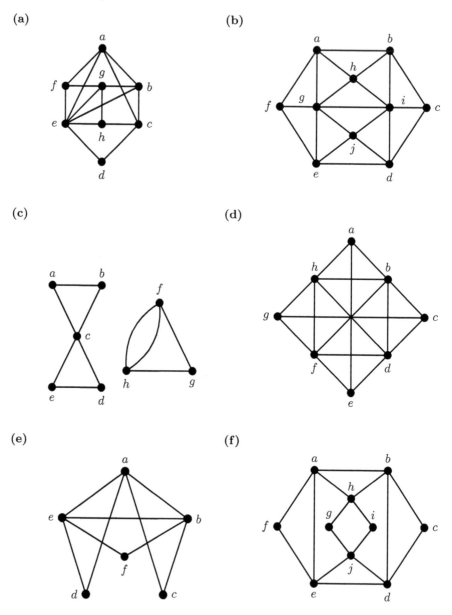

(c)

(d)

(e)

(f)

1.5 For those graphs from Exercise 1.4 that have one, find an Eulerian circuit or Eulerian trail.

1.6 Find an Eulerian circuit or Eulerian trail for each of the graphs below.

(a)   **(b)**

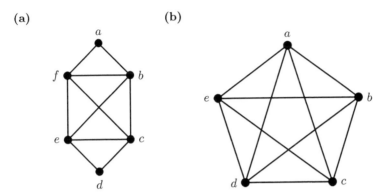

(c)

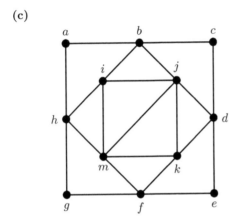

(d)

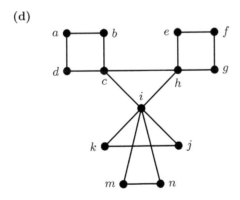

1.7 Find an optimal Eulerization and semi-Eulerization for each of the graphs below.

(a)

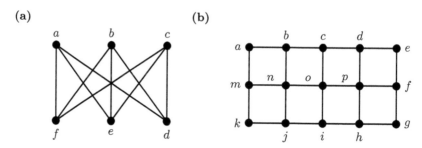

(b)

(c)

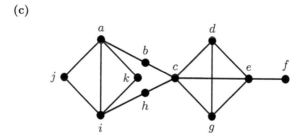

(d)

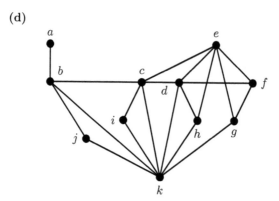

(e)

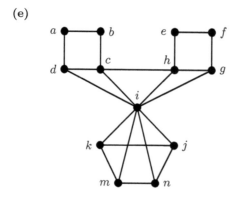

1.8 Find an optimal Eulerization and semi-Eulerization for the graph in Example 1.13 with the edge weights removed.

Use the following map for Problems 1.9 and 1.10.

1.9 Sarah is planning the route for the neighborhood watch night patrol for the community shown in the map above. The person on patrol must walk along each street at least once, including the perimeter of the park.
  (a) Model this scenario as a graph.
  (b) Determine if the graph is Eulerian or semi-Eulerian or neither. Eulerize the graph if it is not Eulerian.
  (c) Find an Eulerian circuit starting and ending at the post office.

1.10 A postal worker is delivering mail along her route shown in the map above. She must walk down both sides of the street if there are houses on both sides and does not need to walk the streets that only border the park (since no houses are in the park).
  (a) Model this scenario as a graph (Hint: make use of multi-edges).
  (b) Determine if the graph is Eulerian or semi-Eulerian or neither. Eulerize the graph if it is not Eulerian.
  (c) Find an Eulerian circuit starting and ending at the post office.

1.11 Repeat Exercise 1.9 with the map below.

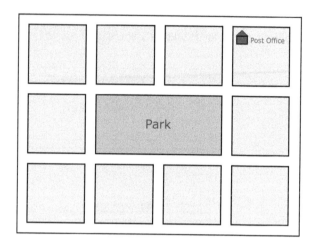

1.12 Repeat Exercise 1.10 with the map above.

1.13 The image below appeared in Euler's original paper on the Königsberg Bridge Problem [14].

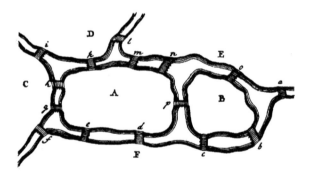

(a) Model the map as a graph (Hint: make use of multi-edges).
(b) Determine if the graph is Eulerian or semi-Eulerian or neither. Eulerize the graph if it is not Eulerian.
(c) Find an Eulerian circuit for either the original graph or its Eulerization.

1.14 Find an optimal Eulerization and semi-Eulerization for each of the weighted graphs below.

(a)

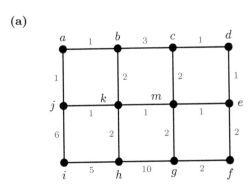

(b)

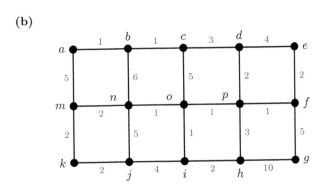

1.15 Find an optimal semi-Eulerization for the graph in Example 1.13 on page 26.

1.16 Fleury's Algorithm can be used to find an Eulerian circuit or an Eulerian trail, yet Hierholzer's is written in such a way as to only work for Eulerian circuits. Modify Hierholzer's Algorithm so it can be used to find an Eulerian trail. (Hint: think about how Fleury's accounts for the two odd vertices.)

1.17 A connected graph G has 8 vertices and 15 edges. You know three vertices have degree 5 and three vertices have degree 3. What are the possible degrees for the remaining two vertices?

Projects

1.18 You have been hired to create a route for the bridge inspector for the city of Pittsburgh, Pennsylvania, which is known for its bridges. The city needs the bridges to be visually inspected on the first day of every month and so needs the route to be as short as possible in order for the inspector to complete his tour in one day. Use a high quality map to create a graph similar to the one that arose from Königsberg. Add edge weights that correspond to distance or time and find an optimal exhaustive route (a tour through the graph that visits each bridge at least once). Write a detailed report for the Pittsburgh city manager outlining your methodology, results, and recommendations for the bridge inspector.

1.19 Pick a city neighborhood. Using a quality map, model the neighborhood as a graph and assign weights to the edges. You may use any logical metric in assigning weights (such as time or distance). Find an optimal route for a street sweeper that must visit each street in the neighborhood.

Chapter 2

Hamiltonian Cycles

Think back to the city of Königsberg. The previous chapter introduced the notion of a circuit and determined when a graph would contain an Eulerian circuit, a special type of circuit that must travel through every edge and vertex. This concept arose from a desire to cross every bridge in the city.

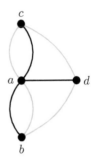

What if we change the requirements ever so slightly so that we are only concerned with the landmasses? This could model a delivery service with customers in every sector of the city. In graph theoretic terms, we are looking for a tour through the graph that hits every vertex exactly once. An example of such a tour on the graph representing Königsberg is shown above. What type of tour is this? If we need to start and end at the same location, we are searching for a cycle. If the starting and ending points can differ, we are searching for a path.

Definition 2.1 A cycle in a graph G that contains every vertex of G is called a *Hamiltonian cycle*. A path that contains every vertex is called a *Hamiltonian path*.

Recall that a cycle or a path can only pass through a vertex once, so the Hamiltonian cycles and paths travel through *every* vertex exactly once.

As with Eulerian circuits, these specific cycles (or paths) are named for the mathematician who first formalized them, Sir William Hamilton. Hamilton posed this idea in 1856 in terms of a puzzle, which he later sold to a game dealer. The "Icosian Game" was a wooden puzzle with numbered ivory pegs where the player was tasked with inserting the pegs so that following them in order would traverse the entire board (shown at the right). Perhaps not too surprisingly, this game was not a big money maker.

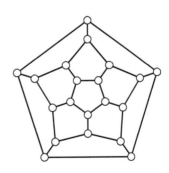

It should be noted that T.P. Kirkman, a contemporary of Hamilton's, did much of the early work in the study of Hamiltonian circuits. Whereas Hamilton primarily focused on one graph, Kirkman was concerned with the existence question: under what conditions will a graph have a Hamiltonian cycle? However, Hamilton deserves credit for publicizing the concept of a cycle that hits every vertex exactly once. This chapter will explore the existence question (when does a graph have a Hamiltonian cycle?), the construction question (how do we find a Hamiltonian cycle?), as well as the optimization question (how do we obtain the best Hamiltonian cycle?).

2.1   Conditions for Existence

When comparing an Eulerian circuit with a Hamiltonian cycle, only one requirement has been lifted: instead of a tour containing every edge and every vertex, we are now only concerned with the vertices. However, as often happens in mathematics, when restrictions are relaxed, the solution either does not exist or finding a solution becomes more difficult.

For the past one hundred fifty years, numerous mathematicians have searched for a solution to the Hamiltonian cycle problem; that is, what are the necessary and sufficient conditions for a graph to contain a Hamiltonian cycle? Recall that a *necessary condition* is a property that must be achieved in order for a solution to be possible and a *sufficient condition* is a property that guarantees the existence of a solution. As we saw in the previous chapter, necessary and sufficient conditions were found for Eulerian circuits — a graph simply needs to be connected with all even vertices in order for an Eulerian circuit to exist. The same is not true for Hamiltonian cycles.

As an initial example, look back at the Königsberg Bridge Problem. The resulting graph does not have an Eulerian circuit but does have a Hamiltonian cycle! The examples below should give some insight into why the difficulty in finding a solution to the Hamiltonian cycle problem occurs.

Example 2.1 Does the graph below have an Eulerian circuit or trail? Hamiltonian cycle or path?

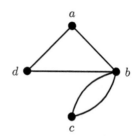

Solution: Since the graph is connected and all vertices are even, we know it has an Eulerian circuit. There is no Hamiltonian cycle since we need to include c in the cycle and by doing so we have already passed through b twice, making it impossible to visit a and d.

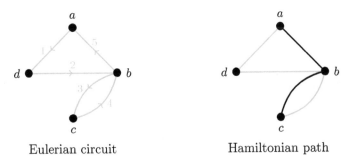

Eulerian circuit   Hamiltonian path

Example 2.2 Does the graph below have an Eulerian circuit or trail? Hamiltonian cycle or path?

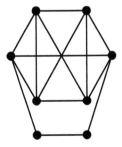

Solution: Since the graph is connected and all vertices are even, we know it is Eulerian. Hamiltonian cycles and Hamiltonian paths also exist. To find one such path, remove any one of the highlighted edges from the Hamiltonian cycle shown below.

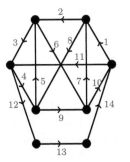

Eulerian circuit

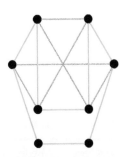

Hamiltonian cycle

Recall from Chapter 1 that if a graph has an Eulerian circuit, then it cannot have an Eulerian trail, and vice versa. The same is not true for the Hamiltonian version:

- If a graph has a Hamiltonian cycle, it automatically has a Hamiltonian path (just leave off the last edge of the cycle to obtain a path).

- If a graph has a Hamiltonian path, it may or may not have a Hamiltonian cycle.

Example 2.3 Does the graph below have an Eulerian circuit or trail? Hamiltonian cycle or path?

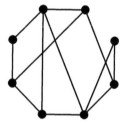

Solution: Since some vertices are odd, we know the graph is not Eulerian. Moreover, since more than two vertices are odd, the graph is not semi-Eulerian. However, this graph does have a Hamiltonian cycle (and so also a Hamiltonian path). Can you find it?

Example 2.4 Does the graph below have an Eulerian circuit or trail? Hamiltonian cycle or path?

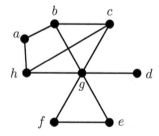

Solution: Since four vertices are odd, we know the graph is neither Eulerian nor semi-Eulerian. This graph does not have a Hamiltonian cycle since d cannot be a part of any cycle. Moreover, this graph does not have a Hamiltonian path since any traversal of every vertex would need to travel through g multiple times.

As should be obvious from the above examples, a few necessary conditions can be placed on the graph to ensure a Hamiltonian cycle is possible. These include, but are not limited to, the following:

(1) G must be connected.

(2) No vertex of G can have degree less than 2.

(3) G cannot contain a **cut-vertex**, that is a vertex whose removal disconnects the graph.

Note that none of the above conditions are sufficient. Look back at the previous examples. Which of them demonstrate this? Can you find other examples?

There are sufficient conditions for a graph to contain a Hamiltonian cycle; but again, although these properties guarantee a Hamiltonian cycle exists for graph satisfying these conditions, not all Hamiltonian graphs must satisfy these properties. We will only include one here for discussion. For further information, see [9].

Theorem 2.2 [12] (Dirac's Theorem) Let G be a graph with $n \geq 3$ vertices. If every vertex of G satisfies $\deg(v) \geq \frac{n}{2}$, then G has a Hamiltonian cycle.

The proof of this theorem boils down to the fact that each vertex has so many edges incident to it that in trying to find a cycle we will never get stuck at a vertex. But this property is *not* a necessary condition.

Example 2.5 Each of the graphs below contains a Hamiltonian cycle, shown in blue. However, only the graph on the left satisfies Dirac's Theorem.

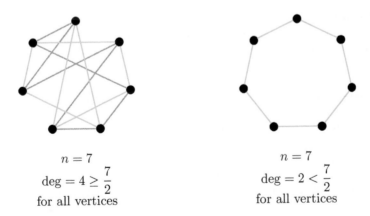

$$n = 7 \qquad\qquad n = 7$$
$$\deg = 4 \geq \frac{7}{2} \qquad\qquad \deg = 2 < \frac{7}{2}$$
$$\text{for all vertices} \qquad\qquad \text{for all vertices}$$

For the remainder of this chapter, we will be focusing not on *if* a graph has a Hamiltonian cycle, but rather how to *find* one in a graph known to have such a cycle. We will mainly focus on a special type of graph, known as a *complete graph*.

Definition 2.3 A simple graph G is **complete** if every pair of distinct vertices is adjacent. The complete graph on n vertices is denoted K_n. The first six complete graphs are shown below.

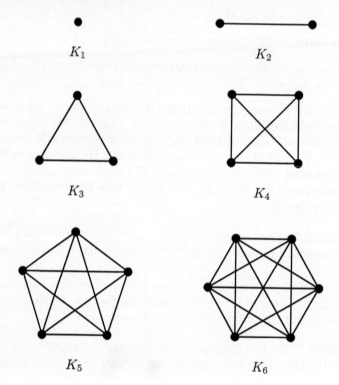

Complete graphs are special for a number of reasons. In particular, if you think of an edge as describing a relationship between two objects, then a complete graph represents a scenario where every pair of vertices satisfies this relationship. Other useful properties of complete graphs are given below.

Properties of K_n

(1) Each vertex in K_n has degree $n - 1$.

(2) K_n has $\dfrac{n(n - 1)}{2}$ edges.

(3) K_n contains the most edges out of all simple graphs on n vertices.

Note that any complete graph (on at least 3 vertices) satisfies the conditions of Dirac's Theorem, and therefore contains a Hamiltonian cycle. In fact, unlike in a general graph, we can easily count the number of distinct Hamiltonian cycles in a complete graph. First we need to define *factorial*.

Definition 2.4 Given a positive integer n, define n *factorial* as

$$n! = n \cdot (n - 1) \cdot (n - 2) \cdots (1)$$

and where we define $0! = 1$.

Factorials are prevalent in many branches of mathematics and appear quite often in counting problems. A few useful properties of factorial arithmetic calculations are shown in the example below. Further computations using factorials can be found in the exercises.

Example 2.6 Compute the following factorials.

- $4! = 4 \cdot 3 \cdot 2 \cdot 1 = 24$

- $5! = 5 \cdot 4 \cdot 3 \cdot 2 \cdot 1 = 5 \cdot 4! = 120$

- $6! = 6 \cdot 5! = 720$

- $\dfrac{6!}{4!} = \dfrac{6 \cdot 5 \cdot (4 \cdot 3 \cdot 2 \cdot 1)}{4!} = \dfrac{6 \cdot 5 \cdot 4!}{4!} = 6 \cdot 5 = 30$

When modeling a problem whose solution is a Hamiltonian cycle (or path) in the appropriate graph, there is often a desired "home" location or starting vertex. We call the starting vertex of a cycle the ***reference point***.

Theorem 2.5 Given a specified reference point, the complete graph K_n has $(n - 1)!$ distinct Hamiltonian cycles. Half of these cycles are reversals of the others.

The idea of the proof is as follows. When starting at the designated vertex there are $n-1$ possible edges to choose from. Once that edge has been traveled to arrive at a new vertex, we cannot pick the edge just traveled and so there remain $n - 2$ edges to choose from. At the next vertex, we cannot travel back to either of the previously chosen vertices, and so there remains $n - 3$ edges available. Continuing in this manner, we have a total of $(n-1) \cdot (n-2) \cdots (2) \cdot (1) = (n - 1)!$ possible Hamiltonian cycles.

2.2 Traveling Salesman Problem

Think back to the question raised at the beginning of this chapter: How should a delivery service plan its route through a city to ensure each customer is reached? Historically, the extensive study of Hamiltonian circuits arose in part from a similar question: A traveling salesman has customers in numerous cities. He must visit each of them and return home, but wishes to do this with the least total cost. Determine the cheapest route possible for the salesman.

In fact, Proctor and Gamble spurred the study of Hamiltonian circuits in the 1960s with a seemingly innocent competition asking for a shortest Hamiltonian circuit visiting 33 cities across the United States. Mathematicians were intrigued and an entire branch of mathematics and computer science developed. For over half a century, some of the brightest minds have tackled the Traveling Salesman Problem (my graph theory professor in college called it "the disease") and numerous books and websites are devoted to finding an optimal solution to both the general question and to specific instances (such as a cycle through all cities in Sweden). A full discussion of the problem is beyond the scope of this book, though you are encouraged to peruse [6] or [7].

The graph that models the general Traveling Salesman Problem is a *weighted complete graph*. Recall from Definition 1.11 that a weighted graph is one in which each edge is assigned a weight, which usually represents either distance, time, or cost. It is standard to use a complete graph since theoretically it should be possible to travel between any two cities.

Brute Force

To find the Hamiltonian cycle of least total weight, one obvious method is to find all possible Hamiltonian cycles and pick the cycle with the smallest total. The method of trying every possibility to find an optimal solution is referred to as an *exhaustive search*, or use of the *Brute Force Algorithm*. This method can be used for any number of problems, not just the Traveling Salesman Problem, though the description below is only for finding Hamiltonian cycles. Knowing this, you might be asking yourself why this problem is still being studied. If we have an algorithm that will find the optimal Hamiltonian cycle, why are mathematicians still interested? The answer will soon become clear.

Brute Force Algorithm

Input: Weighted complete graph K_n.

Steps:

1. Choose a starting vertex, call it v.

2. Find all Hamiltonian cycles starting at v. Calculate the total weight of each cycle.

3. Compare all $(n-1)!$ cycles. Pick one with the least total weight. (Note: there should be at least two options).

Output: Minimum Hamiltonian cycle.

Although both K_3 and K_4 contain Hamiltonian cycles (try it!), the first graph with some complexity is K_5. The example below walks through the process of finding all 24 possible weighted cycles. You are encouraged to find these on your own and then check your solution with the chart provided.

Example 2.7 Liz is planning her next business trip from her hometown of Addison and has determined the cost for travel between any of the five cities she must visit. This information is modeled in the weighted complete graph below, where the weight is given in terms of dollars. Use Brute Force to find all possible routes for her trip.

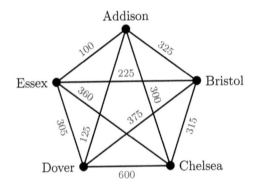

Solution: One method for finding all Hamiltonian cycles, and ensuring you indeed have all 24, is to use alphabetical or lexicographic ordering of the cycles. Note that all cycles must start and end at Addison and we will abbreviate all cities with their first letter. For example, the first cycle is $abcdea$ and appears first in the list below. Below each cycle is its reversal and total weight.

$a\,b\,c\,d\,e\,a$
$a\,e\,d\,c\,b\,a$
1645

$a\,b\,c\,e\,d\,a$
$a\,d\,e\,c\,b\,a$
1430

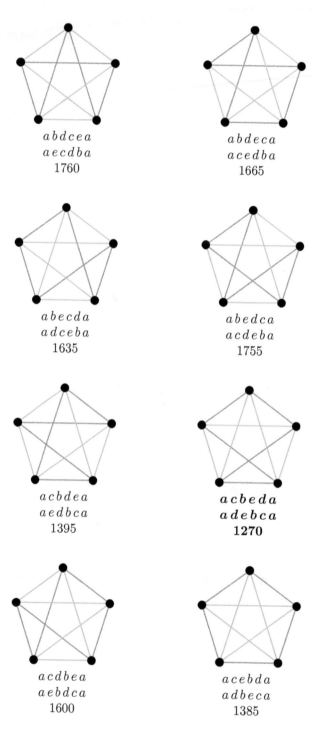

abdcea
aecdba
1760

abdeca
acedba
1665

abecda
adceba
1635

abedca
acdeba
1755

acbdea
aedbca
1395

acbeda
adebca
1270

acdbea
aebdca
1600

acebda
adbeca
1385

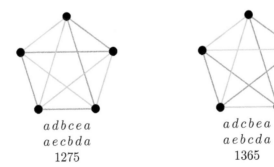

$$a\,d\,b\,c\,e\,a \qquad\qquad a\,d\,c\,b\,e\,a$$
$$a\,e\,c\,b\,d\,a \qquad\qquad a\,e\,b\,c\,d\,a$$
$$1275 \qquad\qquad\qquad 1365$$

From the data above we can identify the optimal cycle as $a\,c\,b\,e\,d\,a$, which provides Liz with the optimal route of Addison to Chelsea to Bristol to Essex to Dover and back to Addison, for a total at a cost of \$1270.

If you attempted to find all 24 cycles by hand, how long did it take you? You may have gotten into a rhythm and improved after the first few, but it still took some time to complete. What if you tried K_6? K_7? K_{15}? How many cycles would you need to check? Ignoring reversals you have 120, 720, and roughly 87 billion cycles to find.

The Traveling Salesman Problem is of interest to mathematicians in part because of how quickly the size of the problem grows. As the size of the input grows (say from K_5 to K_{15}) the number of calculations needed to obtain an output explodes (from 24 to about 87 billion; see Section 7.1 for further discussion). You may be thinking "Yeah...but we have these things called computers that can work faster than any human, so that shouldn't be a problem."

Computer performance is often measured in FLOPS, an acronym for floating-point operations per second. Roughly speaking, a floating-point operation can consist of arithmetic calculations (such as adding or subtracting two numbers). Top of the line desktop processors have performance ratings of roughly 175 billion FLOPS, better known as 175 GFLOPS [23]. At the time of publication, the best supercomputer in the world had a performance rating of about 33 million GFLOPS and the sum of the top 500 supercomputers was 308 million GFLOPS [38].

To determine how quickly these computers would complete the Brute Force Algorithm, we first need to determine the number of FLOPS required. Given a specified starting vertex, we know there are essentially $(n-1)!/2$ possible Hamiltonian cycles for the complete graph K_n. For each of these cycles we need to perform n additions to find its total weight. Once a cycle and its weight have been computed, we must compare them to the previously computed cycle, only keeping in memory the one of least total weight, requiring another $(n-1)!/2$ calculations. Altogether, we estimate the time required to fully implement the

Brute Force Algorithm on K_n is

$$n \cdot \frac{(n-1)!}{2} + \frac{(n-1)!}{2} = \frac{(n+1) \cdot ((n-1)!)}{2}$$

$$= \frac{(n+1)!}{2n} \text{ FLOPS.}$$

Suppose you have been given access to the highest rated supercomputer (and also the top 500) in the world and would like to find the optimal Hamiltonian cycle on K_n for each n from 5 to 50. How long will this take? The table below gives you an estimate of the time requirements for increasing values of n. To put some of these numbers into context, scientists believe the earth is about 4.54 billion years old, that is 4.54×10^9 years! Using Brute Force is computationally impractical for graphs with more than 24 vertices, even when using the top 500 supercomputers in the entire world!

The discussion above illustrates how ineffective Brute Force is when trying to solve an instance of the Traveling Salesman Problem. Although it will eventually find a solution, the time necessary to finish all the computations can be quite unreasonable. Mathematicians have been searching for algorithms that will find the optimal cycle in a relatively short time span; that is, an algorithm that is both efficient and optimal. Not only has no such algorithm been found for the Traveling Salesman Problem, but some mathematicians believe no such algorithm even exists. Further discussion of the difference between efficient and optimal can be found in Section 7.1.

	Supercomputers	
n	**Best**	**Top 500**
5	2×10^{-15} seconds	2×10^{-16} seconds
15	2×10^{-5} seconds	2×10^{-6} seconds
20	40 seconds	4 seconds
21	14 minutes	2 minutes
22	5 hours	32 minutes
23	4.5 days	12 hours
24	16 weeks	12 days
25	7.5 years	10 months
26	2 centuries	2 decades
30	132 million years	14 million years
40	4.1×10^{23} years	4.3×10^{22} years
50	1.4×10^{40} years	1.5×10^{39} years

The next few sections discuss **approximate algorithms,** which are efficient but not optimal. This means they can find a good Hamiltonian cycle without taking too much computational time. In some instances these algorithms may in fact find the optimal cycle, but there is no guarantee that this will always occur. This will become evident through the examples and exercises.

Nearest Neighbor

If you do not have time to run Brute Force, how would you find a good Hamiltonian cycle? You could begin by taking the edge to the "closest" vertex from your starting location, that is the edge of the least weight. And then? Maybe move to the closest vertex from your new location? This strategy is called the *Nearest Neighbor Algorithm*.

Nearest Neighbor Algorithm

Input: Weighted complete graph K_n.

Steps:

1. Choose a starting vertex, call it v. Highlight v.

2. Among all edges incident to v, pick the one with the smallest weight. If two possible choices have the same weight, you may randomly pick one.

3. Highlight the edge and move to its other endpoint u. Highlight u.

4. Repeat steps (2) and (3), where only edges to unhighlighted vertices are considered.

5. Close the cycle by adding the edge to v from the last vertex highlighted. Calculate the total weight.

Output: Hamiltonian cycle.

Example 2.8 Apply the Nearest Neighbor Algorithm to the graph from Example 2.7.

Solution: At each step we will show two copies of the graph. One will indicate the edges under consideration and the other traces the route under construction.

Step 1: The starting vertex is a.

Step 2: The edge of smallest weight incident to a is ae with a weight of 100.

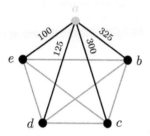
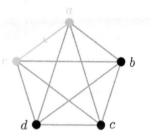

Step 3: From e we only consider edges to b, c or d. Choose edge eb with weight 225.

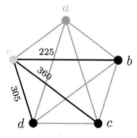
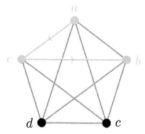

Step 4: From b we consider the edges to c or d. The edge of smallest weight is bc with weight 315.

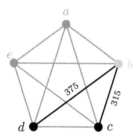
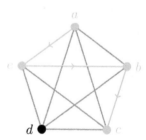

Step 5: Even though cd does not have the smallest weight among all edges incident to c, it is the only choice available.

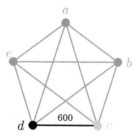
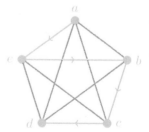

Step 6: Close the circuit by adding da.

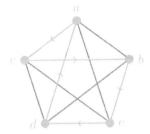

Output: The circuit is $a\,e\,b\,c\,d\,a$ with total weight 1365.

In the example above, the final circuit was Addison to Essex to Bristol to Chelsea to Dover and back to Addison. Although Nearest Neighbor did not find the optimal circuit of total cost \$1270, it did produce a fairly good circuit with total cost \$1365. Perhaps most important was the speed with which Nearest Neighbor found this circuit.

Example 2.8 also illustrates some drawbacks for Nearest Neighbor. First, the last two edges are completely determined since we cannot travel back to vertices that have already been chosen. This could force us to use the heaviest edges in the graph, as happened above. Second, the arbitrary choice of a starting vertex could cause light edges to be eliminated from consideration.

Though we cannot do anything about the former concern, we can address the latter. By using a different starting vertex, the Nearest Neighbor Algorithm may identify a new Hamiltonian cycle, which may be better or worse than the initial cycle. Instead of only considering the circuits starting at the chosen vertex (for example, Liz's hometown of Addison), we will run Nearest Neighbor with each of the vertices as a starting point. This is called **Repetitive Nearest Neighbor**.

Repetitive Nearest Neighbor Algorithm

Input: Weighted complete graph K_n.

Steps:

1. Choose a starting vertex, call it v.

2. Apply the Nearest Neighbor Algorithm.

3. Repeat steps (1) and (2) so each vertex of K_n serves as the starting vertex.

4. Choose the cycle of least total weight. Rewrite it with the desired reference point.

Output: Hamiltonian cycle.

The last step of this algorithm calls for a cycle to be rewritten. Looking back at the table of cycles in Example 2.7, if you did not know the starting vertex was a, you could not discern this from the cycle highlighted in the graph. In fact, any vertex in a cycle could be the reference point, though there is often a designated reference point based on the scenario being modeled.

Example 2.9 Apply the Repetitive Nearest Neighbor Algorithm to the graph from Example 2.7.

Solution: Each of the cycles is shown below, with its original name, the rewritten form with a as the reference point, and the total weight of the cycle. You should notice that the cycle starting at d is the same as the one starting at a, and the cycle starting at b is their reversal. You are encouraged to verify these.

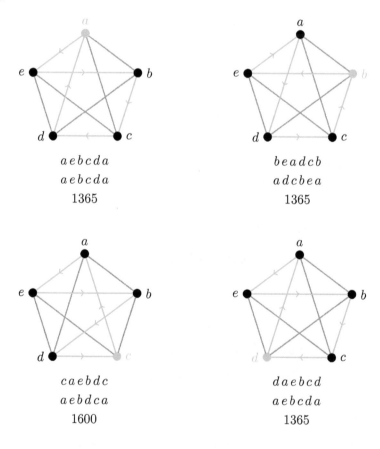

$a\,e\,b\,c\,d\,a$

$a\,e\,b\,c\,d\,a$

1365

$b\,e\,a\,d\,c\,b$

$a\,d\,c\,b\,e\,a$

1365

$c\,a\,e\,b\,d\,c$

$a\,e\,b\,d\,c\,a$

1600

$d\,a\,e\,b\,c\,d$

$a\,e\,b\,c\,d\,a$

1365

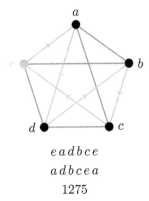

e a d b c e

a d b c e a

1275

It should come as no surprise that Repetitive Nearest Neighbor performs better than Nearest Neighbor; however, there is no guarantee that this improvement will produce the optimal cycle. In the example above, the algorithm found the second best cycle, which only differed from the optimal by $5! This small increase in cost would clearly be worth the time savings over Brute Force.

Cheapest Link

Even though Repetitive Nearest Neighbor addressed the concern of missing small weight edges, it is still possible that some of these will be bypassed as we travel a cycle. The **Cheapest Link Algorithm** attempts to fix this by choosing edges in order of weight as opposed to edges along a tour. Unlike either version of Nearest Neighbor, Cheapest Link does not follow a path that eventually closes into a cycle, but rather chooses edges by weight in such a way that they eventually form a cycle.

Cheapest Link Algorithm

Input: Weighted complete graph K_n.

Steps:

1. Among all edges in the graph pick the one with the smallest weight. If two possible choices have the same weight, you may randomly pick one. Highlight the edge.

2. Repeat step (1) with the added conditions:

   (a) no vertex has three highlighted edges incident to it; and

   (b) no edge is chosen so that a cycle closes before hitting all the vertices.

3. Calculate the total weight.

Output: Hamiltonian cycle.

Example 2.10 Apply the Cheapest Link Algorithm to the graph from Example 2.7.

Solution: In each step shown below, unchosen edges are shown in gray, previously chosen edges are in black, and the newly chosen edge in blue.

Step 1: The smallest weight is 100 for edge *ae*.

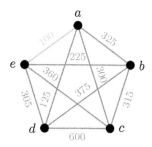

Step 2: The next smallest weight is 125 with edge *ad*.

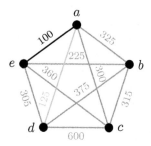

Step 3: The next smallest weight is 225 for *be*. Since this does not close a circuit or cause a vertex to have three incident highlighted edges, we can choose *be*.

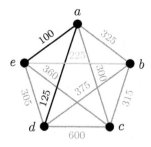

Step 4: Even though *ac* has weight 300, we must bypass it as it would force *a* to have three incident edges that are highlighted. An edge that is skipped will be marked with an X.

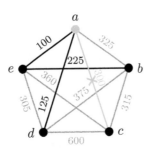

Step 5: The next smallest weight is 305 for edge *ed*, but again we must bypass it as it would close a cycle too early as well as force *e* to have three incident edges that are highlighted.

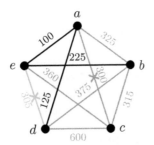

Step 6: The next available is *bc* with weight 315.

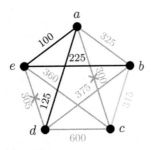

Step 7: At this point, we must close the cycle and there is only one choice *cd*.

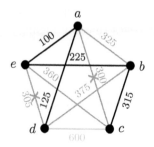

Output: The resulting cycle is $a\,e\,b\,c\,d\,a$ with total weight 1365.

In the example above, Cheapest Link ran into the same trouble as the initial cycle created using Nearest Neighbor; although the lightest edges were chosen, the heaviest also had to be included due to the outcome of the previous steps. However, Cheapest Link will generally perform quite well (in the example above, it found the same cycle as Nearest Neighbor) and is an efficient algorithm.

Nearest Insertion

The downfall of the previous two algorithms is the focus on choosing the smallest weighted edges at every opportunity. In some instances, it may be beneficial to work for a balance — choose more moderate weight edges to avoid later using the heaviest. The **Nearest Insertion Algorithm** does just this by forming a small circuit initially from the smallest weighted edge and balances adding new edges with the removal of a previously chosen edge.

Nearest Insertion Algorithm

Input: Weighted complete graph K_n.

Steps:

1. Among all edges in the graph, pick the one with the smallest weight. If two possible choices have the same weight, you may randomly pick one. Highlight the edge and its endpoints.

2. Pick a vertex that is closest to one of the two already chosen vertices. Highlight the new vertex and its edges to both of the previously chosen vertices.

3. Pick a vertex that is closest to one of the three already chosen vertices. Calculate the increase in weight obtained by adding two new edges and deleting a previously chosen edge. Choose the scenario with the smallest

total. For example, if the cycle obtained from (2) was $a - b - c - a$ and d is the new vertex that is closest to c, we calculate:

$$w(dc) + w(db) - w(cb) \text{ and } w(dc) + w(da) - w(ca)$$

and choose the option that produces the smaller total.

4. Repeat step (3) until all vertices have been included in the cycle.

5. Calculate the total weight.

<u>Output:</u> Hamiltonian cycle.

Example 2.11 Apply the Nearest Insertion Algorithm to the graph from Example 2.7.

Solution: At each step shown below, the graph on the left highlights the edge being added and the graph on the right shows how the cycle is built.

Step 1: The smallest weight edge is ae at 100.

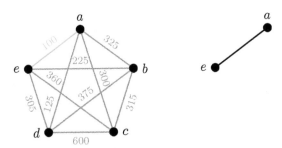

Step 2: The closest vertex to either a or e is d through the edge ad of weight 125. Form a cycle by adding ad and de.

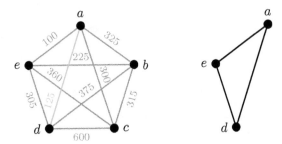

Step 3: The closest vertex to any of a, d or e is b through the edge be with weight 225.

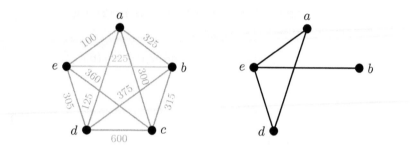

In adding edge be, either ae or de must be removed so that only two edges are incident to e. To determine which is the better choice, compute the following expressions:

$$be + ba - ea = 225 + 325 - 100 = 450$$
$$be + bd - ed = 225 + 375 - 305 = 295$$

Since the second total is smaller, we create a larger cycle by adding edge bd and removing ed.

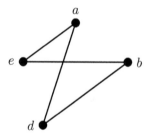

Step 4: The only vertex remaining is c, and the minimum edge to the other vertices is ac with weight 300.

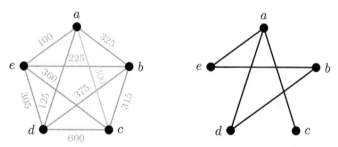

Either ae or ad must be removed. As in the previous step, we compute the following expressions:

$$ca + cd - ad = 300 + 600 - 125 = 775$$
$$ca + ce - ae = 300 + 360 - 100 = 560$$

The second total is again smaller, so we add ce and remove ae.

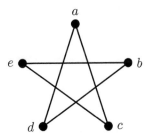

Output: The cycle is $a\,c\,e\,b\,d\,a$ with total weight 1385.

In the example above, Nearest Insertion performed slightly worse than Cheapest Link and Repetitive Nearest Neighbor. Among all three algorithms, Repetitive Nearest Neighbor found the cycle closest to optimal. In general, when comparing algorithm performance we focus less on absolute error ($1275 - 1270 = 5$) but rather on *relative error*.

Definition 2.6 The *relative error* for a solution is given by

$$\epsilon = \frac{Solution - Optimal}{Optimal}$$

Absolute error gives the exact measure away from optimal, but can be misleading if the weights themselves are either very large or very small. Using relative error allows us to compare the performance of an algorithm across multiple examples where the scale may vary. It is important to remember that relative error (and absolute error) can only be calculated when the optimal solution is known.

Example 2.12 Find the relative error for each of the algorithms performed on the graph from Example 2.7.

Solution:

- Repetitive Nearest Neighbor: $\epsilon = \dfrac{1275 - 1270}{1270} = 0.003937 \approx 0.39\%$

- Cheapest Link: $\epsilon = \dfrac{1365 - 1270}{1270} = 0.074803 \approx 7.48\%$

- Nearest Insertion: $\epsilon = \dfrac{1385 - 1270}{1270} = 0.090551 \approx 9.05\%$

It is possible for an approximation algorithm to find the optimal solution.

However, this is unlikely and any of these algorithms could in fact find the absolute worst choice. In addition, no one approximation algorithm will always perform better than the others. There are instances in which one of the algorithms performs better than the others and instances where it performs worse. The next example provides more practice with Cheapest Link and Nearest Insertion. The details of Repetitive Nearest Neighbor for this example can be found in Exercise 2.7.

Example 2.13 Liz needs to head back on the road again but costs have changed, as shown in the table below. Draw a weighted complete graph to model this information. Working directly from the table, find a Hamiltonian cycle using Cheapest Link and Nearest Insertion.

	Addison	Bristol	Chelsea	Dover	Essex
Addison	·	100	350	125	250
Bristol	100	·	265	425	400
Chelsea	350	265	·	300	325
Dover	125	425	300	·	375
Essex	250	400	325	375	·

Solution:

Weighted Graph
To form the weighted graph, each city is represented by a vertex and the cost to travel between them is given by the edge weight. Note that the table above is symmetric. For example traveling from Addison to Bristol (row 1 and column 2) has the same cost as traveling from Bristol to Addison (row 2 and column 1). Thus only one edge is needed between any two vertices.

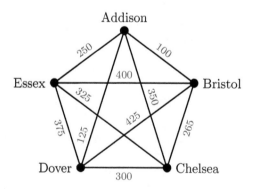

Cheapest Link
Working from the table, we begin with a graph consisting only of the vertices.

At each step we add in the chosen edge as indicated by the algorithm.

Step 1: Begin with edge ab, as it has smallest weight of 100.

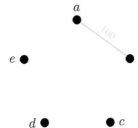

Step 2: The next smallest weight is 125 for edge ad.

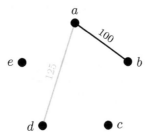

Step 3: Even though ae has weight 250, we must bypass it as it would force a to have three incident blue edges.

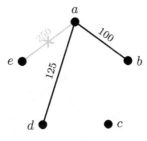

Step 4: The next available is bc of weight 265.

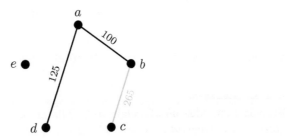

Step 5: The next smallest weight is 300 for edge *cd*, but again we must bypass it as it would close the cycle too early.

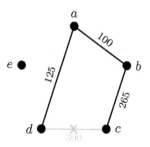

Step 6: The next available is *ce* with weight 325.

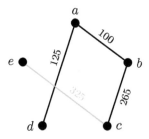

Step 7: At this point, we must close the cycle and there is only one choice *de*.

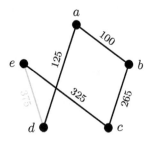

Output: The cycle is *a b c e d a* with total weight 1190 as shown above.

Nearest Insertion

Step 1: Pick edge *ab* of weight 100. The closest vertex to either *a* or *b* is *d* through the edge *ad* of weight 125. Form a cycle by adding *ad* and *bd*.

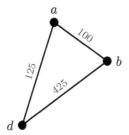

Step 2: The closest vertex to any of a, b or d is e through the edge ae with weight 250.

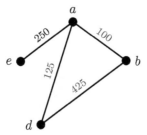

In adding edge ae, either ab or ad must be removed. To determine which is the better choice, compute the following expressions:

$$ae + be - ab = 250 + 400 - 100 = 550$$
$$ae + de - ad = 250 + 375 - 125 = 500$$

Based on the computations above, create a larger cycle by adding edge de and remove ad.

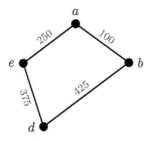

Step 3: The only vertex remaining is c, with the minimum edge of bc.

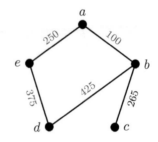

This requires either *ba* or *bd* to be removed. As in the previous step, we compute the following expressions:

$$bc + ca - ba = 265 + 350 - 100 = 515$$
$$bc + cd - bd = 265 + 300 - 425 = 140$$

Choose the addition of *cd* and the removal of *bd*.

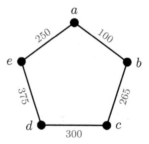

Output: The cycle is *a d c b e a* with total weight of 1290 as shown above.

Note that we cannot calculate the relative error of the cycles obtained in Example 2.13 since we do not know the weight of the optimal cycle. To do so, we would need to apply Brute Force and pick the cycle of minimum total weight.

Hopefully, you now have an idea of why the Traveling Salesman Problem intrigues mathematicians. What appears to be a simple problem is in fact very difficult (time-wise) to solve. There are many more approximation algorithms for the Traveling Salesman Problem (one that is closely related to Nearest Insertion is discussed in Exercise 2.15) and you are encouraged to peruse the website [6] for further discussion of these approaches.

2.3 Digraphs

Thus far, we have restricted our study to symmetric relationships (edge *ab* is the same as edge *ba*). In one sense this accurately models traveling between cities since the distance from Bristol to Chelsea equals the distance from Chelsea to Bristol (think back to the reversals from the application of the Brute Force Algorithm). However, in another sense we are ignoring a real possibility — that traveling between two vertices may not be symmetric. For example, flying from one city to another rarely costs the same in both directions due to customer demand and airport specific fees.

Definition 2.7 A ***directed graph***, or ***digraph***, is a graph $G = (V, A)$ that consists of a vertex set $V(G)$ and an ***arc set*** $A(G)$. An ***arc*** is an ordered pair of vertices.

Example 2.14 Let G be a digraph where $V(G) = \{a, b, c, d\}$ and $A(G) = \{ab, ba, cc, dc, db, da\}$. A drawing of G is given below.

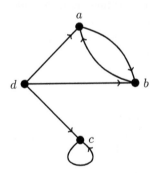

Analogous definitions to those in Definition 1.2 exist for digraphs. A few of these are listed below along with the appropriate references to Example 2.14. Other directed versions of previously defined terminology should be obvious based on context.

Definition 2.8 Let $G = (V, A)$ be a digraph.

- Given an arc xy, the ***head*** is the starting vertex x and the ***tail*** is the ending vertex y.

  - a is the head of arc ab and the tail of arcs da and ba

- Given a vertex x, the ***in-degree*** of x is the number of arcs for which x is a tail, denoted $\deg^-(x)$. The ***out-degree*** of x is the number of arcs for which x is the head, denoted $\deg^+(x)$.

– $\deg^-(a) = 2$, $\deg^-(b) = 2$, $\deg^-(c) = 2$, $\deg^-(d) = 0$
– $\deg^+(a) = 1$, $\deg^+(b) = 1$, $\deg^+(c) = 1$, $\deg^+(d) = 3$

- The **underlying graph** for a digraph is the graph $G' = (V, E)$ which is formed by removing the direction from each arc to form an edge.

- A **directed path** is a path in which the tail of an arc is the head of the next arc in the path.

  – $d \to a \to b$ is a directed path, but $c \to d \to a \to b$ is not since cd is not an arc in the graph.

- A **directed cycle** is a cycle in which the tail of an arc is the head of the next arc in the cycle.

  – $a \to b \to a$ is a directed cycle, but $d \to a \to b \to d$ is not.

Similar to the Handshaking Lemma (Theorem 1.7), the sum of the degrees in a digraph has a relationship to the number of arcs.

Theorem 2.9 Let $G = (V, A)$ be a digraph and $|A|$ denote the number of arcs in G. Then both the sum of the in-degrees of the vertices and the sum of the out-degrees equals the number of arcs; that is, if $V = \{v_1, v_2, \ldots, v_n\}$, then

$$
\begin{aligned}
\deg^-(v_1) + \cdots + \deg^-(v_n) &= |A| \\
&= \deg^+(v_1) + \cdots + \deg^+(v_n)
\end{aligned}
$$

Since each arc will contribute to the in-degree of its tail and to the out-degree of its head, the sum of the in-degrees equals the sum of the out-degrees. Moreover, each arc is counted exactly once in the sum of the in-degrees (or out-degrees) since it has a unique tail (or head).

Recall that determining if a graph has a Hamiltonian cycle or path was quite difficult and no set of properties have been proven to be both necessary and sufficient. It should not come as much of a surprise that finding such paths and cycles in digraphs is even more difficult. In fact, even when the underlying graph is a complete graph, a digraph may not have a Hamiltonian cycle. A few properties should be immediately apparent for a digraph to have a Hamiltonian cycle; namely,

(1) G must be connected.

(2) No vertex of G can have in-degree or out-degree of 0.

(3) G cannot contain a cut-vertex.

The only property listed above that changed from graphs to digraphs was (2). Can you explain the adjustment to considering in-degree and out-degree? What would happen if either of these was 0?

Dirac's Theorem (Theorem 2.2) provided a sufficient condition for a graph to have a Hamiltonian cycle, namely the degree of every vertex must be quite large. A similar result holds for digraphs that again uses in-degree and out-degree.

Theorem 2.10 Let G be a digraph. If $\deg^-(v) \geq \frac{n}{2}$ and $\deg^+(v) \geq \frac{n}{2}$ for every vertex of G, then G has a Hamiltonian cycle.

There has been extensive study of Hamiltonian digraphs; similar to the section on graphs, we will focus on a specific type of digraph known to contain a Hamiltonian cycle. In particular, the Asymmetric Traveling Salesman Problem allows for the weights to differ based upon which direction is taken between vertices.

Asymmetric Traveling Salesman Problem

Consider the problem of the delivery service posed at the beginning of the chapter. The need to visit each customer was translated into the mathematical problem of finding a cycle that contained each vertex of a graph, called a Hamiltonian cycle. When distance or cost was considered, we were searching for the optimal Hamiltonian cycle, referred to as the Traveling Salesman Problem. The Asymmetric Traveling Salesman Problem is to find the optimal Hamiltonian *directed* cycle on a digraph in which the weight of arc ab need not equal the weight of arc ba.

For our purposes, we are only using complete digraphs in which each arc has a positive weight. A **complete digraph** is similar to a complete graph with the added condition that between any two distinct vertices x and y both arcs xy and yx exist in the digraph. This is different from a digraph whose underlying graph is a complete graph, called a *tournament*, since in tournaments exactly one of xy and yx is an arc. Section 7.3 addresses some of the properties of tournaments and their use in mathematical modeling.

Instead of introducing new procedures for finding a Hamiltonian directed cycle on a digraph, we choose to convert a digraph into an undirected graph. This allows us to apply the algorithms from the previous section. Although any of Nearest Neighbor, Cheapest Link, and Nearest Insertion can be used, their implementation will be slightly different from earlier in the chapter due to the nature of the resulting undirected graph.

Undirecting Algorithm

Input: Weighted complete digraph $G = (V, A, w)$.

Steps:

1. For each vertex x make a clone x'. Form the edge xx' with weight 0.

2. For each arc xy form the edge $x'y$.

3. The weight of an edge is equal to the weight of its corresponding arc; that is

- $w(xx') = 0$
- $w(x'y) = w(xy)$
- $w(xy') = w(yx)$.

<u>Output:</u> Weighted clone graph $G' = (V', E', w)$, where V' consists of all vertices from G and their clones and E' the edges described above.

The duplication of vertices allows us to code the weight of an arc into the weight of an edge of an undirected graph. But by adding these clones, we need to ensure that when forming a cycle an arc *into* a vertex is immediately followed by an arc *out* of that same vertex. By giving the edge between a vertex and its clone a weight of 0 (when all other arcs have positive weights), we ensure this edge will always be included in any Hamiltonian cycle. In addition, this extra edge does not impact the total weight of the final cycle.

Example 2.15 Liz is once again heading out on the road to visit her customers, yet this time the direction of a route impacts its cost. This information is displayed below, both in digraph form and as a table. Note that the table is no longer symmetric and an entry is given in terms of row to column; that is, the cost of Addison to Bristol is 325 (row 1 – column 2) and the cost of Bristol to Addison is 375 (row 2 – column 1). Apply the Undirecting Algorithm to the digraph to get its weighted clone graph.

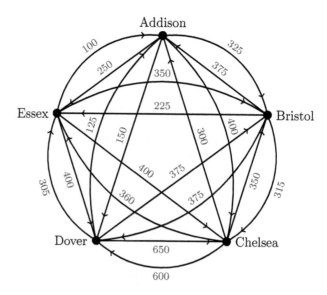

	Addison	Bristol	Chelsea	Dover	Essex
Addison	·	325	400	150	250
Bristol	375	·	315	375	225
Chelsea	300	350	·	600	360
Dover	125	375	650	·	305
Essex	100	350	400	400	·

Solution: Both the weighted clone graph and table of edge weights are given below (with vertex names in place of cities). Note that the new table is symmetric, with a copy of the original table in the lower left quadrant and its mirror image in the upper right quadrant. Edges that do not exist are indicated by a dot in the table.

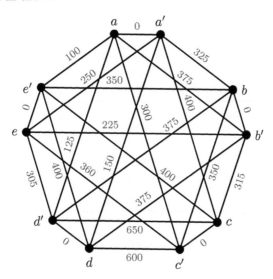

	a	b	c	d	e	a'	b'	c'	d'	e'
a	·	·	·	·	·	0	375	300	125	100
b	·	·	·	·	·	325	0	350	375	350
c	·	·	·	·	·	400	315	0	650	400
d	·	·	·	·	·	150	375	600	0	400
e	·	·	·	·	·	250	225	360	305	0
a'	0	325	400	150	250	·	·	·	·	·
b'	375	0	315	375	225	·	·	·	·	·
c'	300	350	0	600	360	·	·	·	·	·
d'	125	375	650	0	305	·	·	·	·	·
e'	100	350	400	400	0	·	·	·	·	·

Note that even though the input of the algorithm above is a complete digraph, the output is *not* a complete graph (which edges are missing?). The Undirecting Algorithm can be applied to a digraph that is not complete; however, the resulting graph may not have a Hamiltonian cycle. When applied to a complete digraph G on n vertices, the resulting graph G' is guaranteed to have a Hamiltonian cycle since the conditions of Dirac's Theorem are satisfied (G' has $2n$ vertices, each of which has degree n). The cycle from G' can then be translated to a Hamiltonian directed cycle of the digraph G where vertex copies get reduced back to a single vertex. Any cycle in a weighted clone graph will alternate between original vertices from the digraph and their clones.

Example 2.16 Apply Nearest Neighbor to the graph from Example 2.15 with starting vertices a, a', e and e'. Translate each of these cycles into its directed cycles.

Solution: The table below lists the four cycles found using Nearest Neighbor and their conversion to a directed cycle in the digraph. Note that cycles beginning with a clone vertex must be reversed in the translation back into a direct cycle.

Nearest Neighbor Cycle	Conversion	Total Weight
$a\,a'\,d\,d'\,e\,e'\,b\,b'\,c\,c'\,a$	$a \to d \to e \to b \to c \to a$	1420
$a'\,a\,e'\,e\,b'\,b\,c'\,c\,d'\,d\,a'$	$a \to d \to c \to b \to e \to a$	1475
$e\,e'\,a\,a'\,d\,d'\,b\,b'\,c\,c'\,e$	$e \to a \to d \to b \to c \to e$	1300
$e'\,e\,b'\,b\,c'\,c\,a'\,a\,d'\,d\,e'$	$e \to d \to a \to c \to b \to e$	1500

A drawing for the cycle beginning at a' is shown below, as well as its conversion in the digraph.

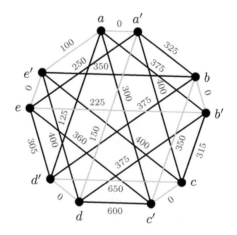

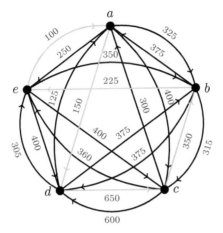

Unlike in the symmetric case, reversals of a Hamiltonian cycle in a digraph (that is not a complete digraph) might not exist, and those that do will most likely result in a different total weight. Thus when applying the Nearest Neighbor Algorithm to graphs formed using the Undirecting Algorithms, we must consider starting at both copies of a vertex, such as a and a' shown above.

Cheapest Link or Nearest Insertion Algorithms create new challenges when applied to a graph formed using the Undirecting Algorithm. In particular, all edges of weight 0 must be included in the final cycle. This happens naturally for Cheapest Link since picking the edges of minimum weight would initially result in choosing all the weight 0 edges as no two of these are adjacent (so there is no concern of a vertex having degree 3 or closing the circuit too early). The example below shows how to adjust the Cheapest Link Algorithm for use on the weighted clone graph.

Example 2.17 Apply Cheapest Link to the graph from Example 2.15. Translate the cycle into the directed cycle, find its total weight, and rewrite it for a reference point of Addison.

Solution:

Step 1: Since all edges between a vertex and its clone have a weight of 0, all of these edges will be chosen. We condense these into one step.

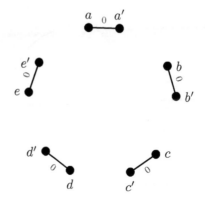

Step 2: The edge of smallest weight is ae' at 100. This is added to the graph below.

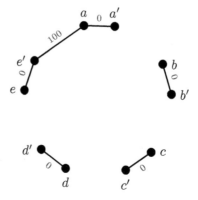

Step 3: The next lowest is 125 for ad'; however, we cannot choose this edge since it would cause a to be incident to three highlighted edges. We skip this edge and move to $a'd$, which has a weight of 150.

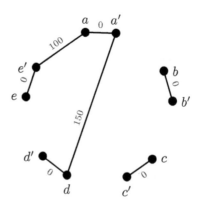

Step 4: The next lowest edge weight is 225 for $b'e$. This is a valid choice and added to the graph below.

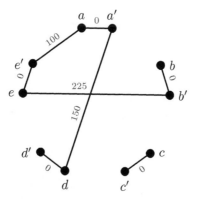

Step 5: The next five smallest weights come from ineligible edges (try it!). The next smallest valid edge is $c'b$ with weight is 350.

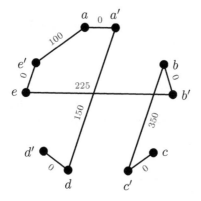

Step 6: To close the circuit, we must choose cd' with weight 650. The resulting circuit in the weighted clone graph is $a\,a'\,d\,d'\,c\,c'\,b\,b'\,e\,e'\,a$.

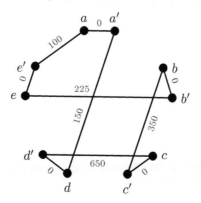

Step 7: We now convert the cycle from the previous step into the digraph for the original digraph. This gives the directed cycle $a \to d \to c \to b \to e \to a$ with total weight 1640.

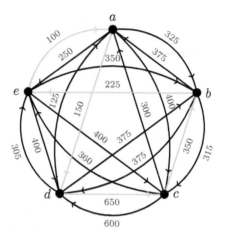

Unlike Cheapest Link, the Nearest Insertion Algorithm in its original form does not function on the weighted clone graphs from this section. The main concern is there are no cycles of length 3 in the graph created by the Undirecting Algorithm (see Exercise 2.13) and the second step of Nearest Insertion results in a cycle on three vertices originating from the cheapest edge of the graph. A modification of Nearest Insertion for the weighted clone graphs, which treats the weight 0 edges differently, appears in Exercise 2.14.

2.4   Exercises

2.1 Compute the following factorials:
(a) 8!
(b) 12!
(c) 16!

2.2 Simplify the following factorials:
(a) $9 * 8!$
(b) $\dfrac{11!}{8!}$
(c) $6! * \dfrac{7!}{5!}$

2.3 How many different Hamiltonian cycles are there for K_4? K_8? K_{10}? Draw all possible Hamiltonian cycles for K_4.

2.4 Find a solution to the Icosian Game shown on page 35.

2.5 Find a Hamiltonian cycle for the graph in Example 2.3.

2.6 For each of the graphs below, determine if G

    (i) definitely has a Hamiltonian cycle;

    (ii) definitely does not have a Hamiltonian cycle; or

    (iii) may or may not have a Hamiltonian cycle.

Explain your answer.

    (a) G has vertices of degree $3, 3, 3, 4, 4, 5$.

    (b) G is connected with 10 vertices, all of which have degree 6.

    (c) G has vertices of degree $1, 2, 2, 3, 5, 5$.

    (d) G is connected with vertices of degree $2, 2, 3, 3, 4, 4$.

    (e) G has vertices of degree $0, 2, 2, 4, 4, 5, 5$.

2.7 Apply Repetitive Nearest Neighbor to the graph from Example 2.13.

2.8 Find a Hamiltonian cycle for each of the graphs below using (i) Repetitive Nearest Neighbor (ii) Cheapest Link and (iii) Nearest Insertion.

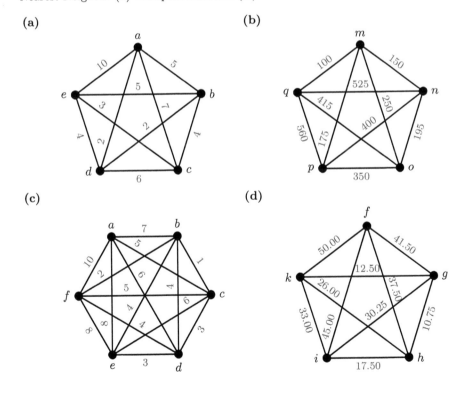

(e)

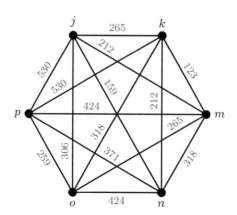

2.9 Chris wants to visit his 4 brothers over the holidays and has determined the costs as shown in the table below. Find a route (and its total weight) for Chris using
    (a) Repetitive Nearest Neighbor
    (b) Cheapest Link
    (c) Nearest Insertion

	Chris	David	Evan	Frank	George
Chris	·	325	300	125	100
David	325	·	215	375	225
Evan	300	215	·	400	275
Frank	125	375	400	·	305
George	100	225	275	305	·

2.10 June and Tori are planning their annual winery tour of Virginia. They want to plan their route so they can see as many of the wineries in one day as possible and this year will be staying at the inn at Mt. Eagle Winery. The chart below lists the wineries and the time (in minutes) between each one. Find a possible route (and its total time) for June and Tori using
    (a) Repetitive Nearest Neighbor
    (b) Cheapest Link
    (c) Nearest Insertion
    (d) and determine if they can visit all six locations in one day.

	Bluebird Wines	Cardinal Winery	Elk Point Vineyard	Red Fox Wines	Graybird Vineyard	Mt. Eagle Winery
Bluebird	·	41	58	43	51	49
Cardinal	41	·	60	7	62	33
Elk Point	58	60	·	75	67	53
Red Fox	43	7	75	·	64	36
Graybird	51	62	67	64	·	68
Mt. Eagle	49	33	53	36	68	·

2.11 Using the digraph below,

(a) Apply the Undirecting Algorithm to find the weighted clone graph.

(b) Using your result from (a), apply the Nearest Neighbor Algorithm with starting vertices a, a', c and c' and convert your results to directed cycles in the digraph. Find the total weight of each directed cycle.

(c) Using your result from (a), apply the Cheapest Link Algorithm and convert your result to a directed cycle in the digraph and find its total weight.

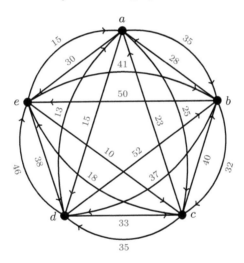

2.12 Leena will be visiting her clients around Europe for the month of April. She has tried to estimate the cost of travel between two cities, using various modes of transportation and discovered the cost depends on the direction of travel. The table below gives these estimates (similar to that of Example 2.15).

(a) Draw the directed graph representing the information in the chart below.

(b) Apply the Undirecting Algorithm to find the weighted clone graph.

(c) Using your result from (a), apply the Nearest Neighbor Algorithm with starting vertices a, a', d and d' and convert your results to directed cycles in the digraph. Find the total weight of each directed cycle.

(d) Using your result from (a), apply the Cheapest Link Algorithm and convert your result to a directed cycle in the digraph and find its total weight.

	Amsterdam	Bern	Düsseldorf	Genoa	Munich
Amsterdam	·	415	375	280	300
Bern	500	·	425	110	250
Düsseldorf	300	425	·	375	240
Genoa	150	200	500	·	400
Munich	275	350	315	400	·

2.13 Explain why no cycles of length three exist in the graph resulting from applying the Undirecing Algorithm to a complete digraph.

2.14 Determine a modification of Nearest Insertion that will allow it to be used on a graph obtained from complete digraph using the Undirecting Algorithm. Use your modification on the graph from Example 2.15. (Hint: the initial cycle should start from the lowest nonzero edge and should have length 4.)

Projects

2.15 The Nearest Insertion Algorithm finds a Hamiltonian cycle by expanding smaller cycles through the addition of the closest vertex to that cycle. It suffers from the same problem as the other algorithms in that a large edge may be chosen in the last step of the algorithm. A variation, called **Farthest Insertion**, first considers the vertices farthest apart since any Hamiltonian cycle must include both of them. In doing so, later additions of vertices will either reduce the cycle weight or increase it by small margins. The description of the algorithm appears below.

Farthest Insertion Algorithm

Input: Weighted complete graph $G = (V, E)$.

Steps:

1. Pick a starting vertex v_1.

2. Choose the vertex v_2 that has the highest weighted edge to v_1.

3. Form a list $(w_1, w_2, w_3, \ldots, w_n)$ where the entry in location i is the minimum weighted edge from v_i to either of v_1 and v_2. The entries for v_1 and v_2 will be left blank (denoted by $-$).

4. Choose vertex x with the largest value from the list created in Step 3. Form the cycle $v_1\, v_2\, x\, v_1$.

5. Update the list from Step 3 so the entries are now the weights from the chosen to unchosen vertices. Choose the next vertex y with largest value in the list.

6. Append the cycle of chosen vertices with y by removing one of the edges from that cycle. Determine which edges to add and subtract by choosing

the lowest total as in Nearest Insertion; that is, if the cycle obtained from Step 4 was $a - b - c - a$ and d is the new vertex to add along with edge to dc, we calculate:

$$w(dc) + w(db) - w(cb) \text{ and } w(dc) + w(da) - w(ca)$$

and choose the option that produces the smaller total.

7. Repeat Steps (5) and (6) until all vertices have been included in the cycle.

Output: Hamiltonian cycle.

Apply Farthest Insertion to the graphs from Examples 2.7 and 2.13.

2.16 Each morning a collection of online orders arrives at the warehouse of a large retailer. Steve, the warehouse manager, must ensure the items are packed and put onto the truck for shipment. However, the items are different every morning and are located in varying locations in the large warehouse. Steve has come to you for help in determining the best method for pulling stock from the shelves. Write a report detailing the Traveling Salesman Problem and how it applies to the warehouse. As part of your report, determine a route for the items shown in the map below. The route must start and end at the packaging bay (p) and the time required for moving down a long aisle is 45 seconds and down a short aisle or between aisles is 10 seconds. For example, it takes 85 seconds to get from item a to item b since four short segments and one long segment are used. Include a weighted graph and discussion of which algorithm(s) you used and if your route is known to be optimal.

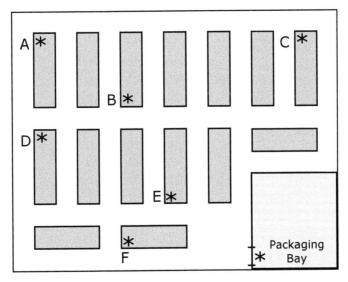

2.17 Come up with a business that needs to solve a Traveling Salesman Problem. Name the business and describe why they are working on this problem. Make sure to include a good reason why finding a Hamiltonian cycle is necessary for their business. The table on the next two pages lists the distance between the top 25 cities based on their population in 2014. Choose 6 cities, draw the weighted graph, and apply the algorithms from this chapter to determine a good Hamiltonian cycle. Discuss the advantages and disadvantages of each technique and if you know any of your cycles is optimal.

Top 25 U.S. Cities by Population

		NYC	LAX	CHI	HOU	PHI	PHX	SAT	SND	DAL	SJC	AUS	JAX	SFO
New York	(NYC)	*	2448	712	1419	81	2142	1583	2431	1372	2552	1513	836	2569
Los Angeles	(LAX)	2448	*	1744	1372	2391	357	1203	112	1239	306	1226	2146	348
Chicago	(CHI)	712	1744	*	943	664	1453	1054	1733	806	1840	981	865	1857
Houston	(HOU)	1419	1372	943	*	1341	1015	189	1303	225	1610	146	822	1644
Philadelphia	(PHI)	81	2391	664	1341	*	2080	1507	2370	1299	2502	1437	759	2520
Phoenix	(PHX)	2142	357	1453	1015	2080	*	848	299	886	615	869	1793	654
San Antonio	(SAT)	1583	1203	1054	189	1507	848	*	1128	253	1453	74	1011	1489
San Diego	(SND)	2431	112	1733	1303	2370	299	1128	*	1183	417	1156	2090	459
Dallas	(DAL)	1372	1239	806	225	1299	886	253	1183	*	1451	182	908	1483
San Jose	(SJC)	2552	306	1840	1610	2502	615	1453	417	1451	*	1466	2344	42
Austin	(AUS)	1513	1226	981	146	1437	869	74	1156	182	1466	*	960	1502
Jacksonville	(JAX)	836	2146	865	822	759	1793	1011	2090	908	2344	960	*	2373
San Francisco	(SFO)	2569	348	1857	1644	2520	654	1489	459	1483	42	1502	2373	*
Indianapolis	(IND)	644	1808	165	822	583	1498	1001	1787	764	1927	927	700	1947
Columbus	(COL)	477	1976	276	993	415	1666	1141	1955	914	2092	1068	670	2111
Fort Worth	(FTW)	1400	1209	825	237	1327	855	241	1152	31	1422	174	938	1454
Charlotte	(CLT)	532	2117	588	927	451	1781	1105	2079	929	2275	1040	342	2299
Detroit	(DET)	481	1982	237	1107	442	1689	1240	1971	1000	2073	1166	834	2089
El Paso	(ELP)	1899	703	1250	672	1831	560	500	630	568	959	526	1469	996
Seattle	(SEA)	2405	961	1735	1891	2376	1116	1788	1065	1682	711	1772	2454	680
Denver	(DEN)	1629	831	919	879	1577	586	803	834	663	928	773	1467	948
Washington D.C.	(WDC)	204	2297	594	1220	123	1980	1387	2272	1183	2418	1318	649	2438
Memphis	(MEM)	954	1602	483	485	880	1261	633	1560	420	1774	561	591	1801
Boston	(BOS)	190	2594	850	1606	271	2298	1767	2582	1551	2681	1696	1018	2696
Nashville	(NSH)	759	1779	398	666	684	1445	824	1742	617	1936	753	500	1961

(continued on the next page)

Top 25 U.S. Cities by Population

		IND	COL	FTW	CLT	DET	ELP	SEA	DEN	WDC	MEM	BOS	NSH
New York	(NYC)	644	477	1400	532	481	1899	2405	1629	204	954	190	759
Los Angeles	(LAX)	1808	1976	1209	2117	1982	703	961	831	2297	1602	2594	1779
Chicago	(CHI)	165	276	825	588	237	1250	1735	919	594	483	850	398
Houston	(HOU)	867	993	237	927	1107	672	1891	879	1220	485	1606	666
Philadelphia	(PHI)	583	415	1327	451	442	1831	2376	1577	123	880	271	684
Phoenix	(PHX)	1498	1666	855	1781	1689	560	1116	586	1980	1261	2298	1445
San Antonio	(SAT)	1001	1141	241	1105	1240	500	1788	803	1387	633	1767	824
San Diego	(SND)	1787	1955	1152	2079	1971	630	1065	834	2272	1560	2582	1742
Dallas	(DAL)	764	914	31	929	1000	568	1682	663	1183	420	1551	617
San Jose	(SJC)	1927	2092	1422	2275	2073	959	711	928	2418	1774	2681	1936
Austin	(AUS)	927	1068	174	1040	1166	526	1772	773	1318	561	1696	753
Jacksonville	(JAX)	700	670	938	342	834	1469	2454	1467	649	591	1018	500
San Francisco	(SFO)	1947	2111	1454	2299	2089	996	680	948	2438	1801	2696	1961
Indianapolis	(IND)	*	168	789	428	240	1261	1870	1000	491	384	806	252
Columbus	(COL)	168	*	939	348	164	1425	2011	1165	327	510	643	334
Fort Worth	(FTW)	789	939	*	960	1023	537	1661	644	1212	449	1578	646
Charlotte	(CLT)	428	348	960	*	506	1491	2283	1357	330	520	721	340
Detroit	(DET)	240	164	1023	506	*	1476	1935	1156	394	624	612	471
El Paso	(ELP)	1261	1425	537	1491	1476	*	1378	557	1722	973	2068	1166
Seattle	(SEA)	1870	2011	1661	2283	1935	1378	*	1021	2324	1866	2488	1973
Denver	(DEN)	1000	1165	644	1357	1156	557	1021	*	1491	878	1767	1022
Washington D.C.	(WDC)	491	327	1212	330	394	1722	2324	1491	*	763	394	567
Memphis	(MEM)	384	510	449	520	624	973	1866	878	763	*	1136	197
Boston	(BOS)	806	643	1578	721	612	2068	2488	1767	394	1136	*	943
Nashville	(NSH)	252	334	646	340	471	1166	1973	1022	567	197	943	*

Chapter 3

Paths

The previous two chapters focused on exhaustive routes through a graph. Whether the objective was to visit vertices or edges, in both cases we were concerned with including all of them. Compare that to the following scenario:

> Pamela is driving from Bennington to Brattleboro and needs to do so as quickly as possible (without speeding of course!). She has estimates of how long each portion of the trip takes, but there are multiple routes available.

If we drew the graph that models the possible routes, with weights assigned to the edges representing expected time, would Pamela be interested in an Eulerian circuit or trail? Hamiltonian cycle or path? No! The former would mean she takes every possible road, whereas the latter would represent going through every possible intersection. Neither of these is the correct graph model for her scenario. So then, what graph model solves this route problem?

The following section will discuss a method for finding a shortest route within a graph, as well as revisit the Chinese Postman Problem from Chapter 1. In addition, a method for determining a work schedule for various interdependent processes provides another application of routes within a digraph.

3.1 Shortest Paths

The problem above can be described in graph theoretic terms as the search for a *shortest path* on a weighted graph. Recall that a path is a sequence of vertices in which there is an edge between consecutive vertices and no vertex is repeated. As with the algorithms for the Traveling Salesman Problem, the weight associated to an edge may represent more than just distance (e.g., cost or time) and the shortest path really indicates the path of least total weight. Note, in this section we will only investigate the construction question (how to find a shortest path) since the existence question (does a shortest path exist) is quickly answered by simply knowing if the graph is connected.

In 1956 Edsger W. Dijkstra proposed the algorithm we are about to study not out of necessity for finding a shortest route, but rather as a demonstration of the power of a new "automatic computer" at the Mathematical Centre in

Amsterdam. The goal was to have a question easily understood by a general audience while also allowing for audience participation in determining the inputs of the algorithm. In Dijkstra's own words "the demonstration was a great success." [11] Perhaps more surprising is how important this algorithm would become to modern society — almost every GIS (Geographic Information System, or mapping software) uses a modification of Dijkstra's algorithm to provide directions. In addition, Dijkstra's algorithm provides the backbone of many routing systems and some studies in epidemiology.

Dijkstra's Algorithm

Numerous versions of Dijkstra's Algorithm exist, though two basic descriptions adhere to Dijkstra's original design. In one, the shortest paths from your chosen vertex to all other vertices are found. Though useful in its own right, this does not properly model the question we are trying to answer, namely what is the shortest route from point a to (a specific) point b (or in Pamela's case, the shortest route from Bennington to Brattleboro). The version of Dijkstra's Algorithm shown below stops once the required ending vertex has been reached and is more faithful to the format in the original publication (see [10]). For more discussion of the lengthier version, see Exercise 3.2 or [40].

Dijkstra's Algorithm is a bit more complex than the algorithms we have studied so far. Each vertex is given a two-part label $L(v) = (x, (w(v))$. The first portion of the label is the name of the vertex used to travel to v. The second part is the weight of the path that was used to get to v from the designated starting vertex. At each stage of the algorithm, we will consider a set of *free* vertices, denoted by an F below. Free vertices are the neighbors of previously visited vertices that are themselves not yet visited.

Dijkstra's Algorithm

Input: Weighted connected simple graph $G = (V, E)$ and vertices designated as *Start* and *End*.

Steps:

1. For each vertex x of G, assign a label $L(x)$ so that $L(x) = (-, 0)$ if $x = Start$ and $L(x) = (-, \infty)$ otherwise. Highlight *Start*.

2. Let $u = Start$ and define F to be the neighbors of u. Update the labels for each vertex v in F as follows:

   if $w(u) + w(uv) < w(v)$, then redefine $L(v) = (u, w(u) + w(uv))$

   otherwise do not change $L(v)$

3. Highlight the vertex with lowest weight as well as the edge uv used to update the label. Redefine $u = v$.

4. Repeat steps (2) and (3) until the ending vertex has been reached. In all future iterations, F consists of the un-highlighted neighbors of all previously highlighted vertices and the labels are updated only for those vertices that are adjacent to the last vertex that was highlighted.

5. The shortest path from *Start* to *End* is found by tracing back from *End* using the first component of the labels. The total weight of the path is the weight for *End* given in the second component of its label.

Output: Highlighted path from *Start* to *End* and total weight $w(End)$.

Perhaps the most complex portion of this algorithm is the labeling of the vertices and how they are updated with iterations of Step (2) and Step (3). In the initial step of Dijkstra's Algorithm, all vertices have no entry in the first part of the label and the second part is 0 for the starting vertex and ∞ for all others. Note that the set F of free vertices consists of all neighbors of highlighted vertices and all are under consideration for becoming the next highlighted vertex. It is important that we do not only consider the neighbors of the last vertex highlighted, as a path from a previously chosen vertex may in fact lead to the shortest path. The example below provides a detailed explanation in the updating of the vertex labels and how to use them to find a shortest path.

Example 3.1 Apply Dijkstra's Algorithm to the graph below where $Start = g$ and $End = c$.

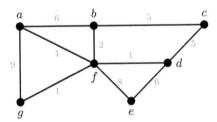

Solution: In each step, the label of a vertex will be shown as a subscript.

Step 1: Highlight g. Define $L(g) = (-,0)$ and $L(x) = (-,\infty)$ for all $x = a, \cdots, f$.

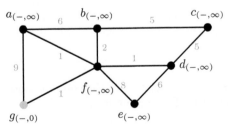

Step 2: Let $u = g$. Then the neighbors of g comprise $F = \{a, f\}$. We compute

$$w(g) + w(ga) = 0 + 9 = 9 < \infty = w(a)$$
$$w(g) + w(gf) = 0 + 1 = 1 < \infty = w(f)$$

Update $L(a) = (g, 9)$ and $L(f) = (g, 1)$. Since the minimum weight for all vertices in F is that of f, we highlight the edge gf and the vertex f.

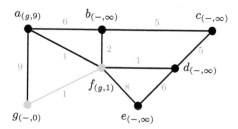

Step 3: Let $u = f$. Then the neighbors of all highlighted vertices are $F = \{a, b, d, e\}$. We compute

$$w(f) + w(fa) = 1 + 1 = 2 < 9 = w(a)$$
$$w(f) + w(fb) = 1 + 2 = 3 < \infty = w(b)$$
$$w(f) + w(fd) = 1 + 1 = 2 < \infty = w(d)$$
$$w(f) + w(fe) = 1 + 8 = 9 < \infty = w(e)$$

Update $L(a) = (f, 2)$, $L(b) = (f, 3)$, $L(d) = (f, 2)$ and $L(e) = (f, 9)$. Since the minimum weight for all vertices in F is that of a or d, we choose to highlight the edge fa and the vertex a.

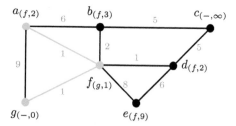

Step 4: Let $u = a$. Then the neighbors of all highlighted vertices are $F = \{b, d, e\}$. Note, we only consider updating the label for b since this is the only vertex adjacent to a, the vertex highlighted in the previous step.

$$w(a) + w(ba) = 2 + 6 = 8 \not< 2 = w(b)$$

We do not update the label for b since the computation above is not less than the current weight of b. The minimum weight for all vertices in F is that of d, and so we highlight the edge fd and the vertex d.

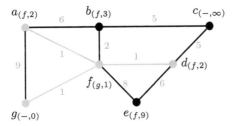

Step 5: Let $u = d$. Then the neighbors of all highlighted vertices are $F = \{b, c, e\}$. We compute

$$w(d) + w(dc) = 2 + 5 = 7 < \infty = w(c)$$
$$w(d) + w(de) = 2 + 6 = 8 < 9 = w(e)$$

Update $L(c) = (d, 7)$ and $L(e) = (d, 8)$. Since the minimum weight for all vertices in F is that of b, we highlight the edge bf and the vertex b.

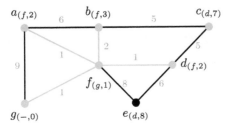

Step 6: Let $u = b$. Then the neighbors of all highlighted vertices are $F = \{c, e\}$. However, we only consider updating the label of c since e is not adjacent to b. Since

$$w(b) + w(bc) = 3 + 5 = 8 \not< 7 = w(c)$$

we do not update the labels of any vertices. Since the minimum weight for all vertices in F is that of c we highlight the edge dc and the vertex c. This terminates the iterations of the algorithm since our ending vertex has been reached.

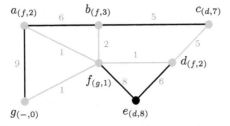

Output: The shortest path from g to c is $g\,f\,d\,c$, shown in blue above. This path has a total weight 7, as shown by the label of c.

The example below is a minor modification of the one above. By simply changing the weight of one edge, the algorithm performs differently and we find a change in the shortest path. This should further demonstrate the need to consider all neighbors of the highlighted vertices, not just the last one highlighted.

Example 3.2 Apply Dijkstra's Algorithm to the graph below where $Start = g$ and $End = c$.

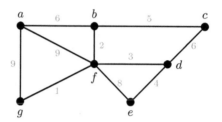

Solution: As in Example 3.1, the label of a vertex will be shown as a subscript.

Step 1: Highlight g. Define $L(g) = (-,0)$ and $L(x) = (-,\infty)$ for all $x = a, \cdots, f$.

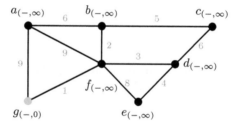

Step 2: Let $u = g$. Then the neighbors of g comprise $F = \{a, f\}$. We compute

$$w(g) + w(ga) = 0 + 9 = 9 < \infty = w(a)$$
$$w(g) + w(gf) = 0 + 1 = 1 < \infty = w(f)$$

Update $L(a) = (g,9)$ and $L(f) = (g,1)$. Since the minimum weight for all vertices in F is that of f, we highlight the edge gf and the vertex f.

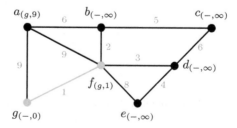

Step 3: Let $u = f$. Then the neighbors of all highlighted vertices are $F = \{a, b, d, e\}$. We compute

$$w(f) + w(fa) = 1 + 9 = 10 \not< 9 = w(a)$$
$$w(f) + w(fb) = 1 + 2 = 3 < \infty = w(b)$$
$$w(f) + w(fd) = 1 + 3 = 4 < \infty = w(d)$$
$$w(f) + w(fe) = 1 + 8 = 9 < \infty = w(e)$$

Update $L(b) = (f, 3)$, $L(d) = (f, 4)$ and $L(e) = (f, 9)$. Since the minimum weight for all vertices in F is that of b, we highlight the edge fb and the vertex b.

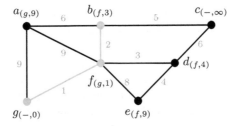

Step 4: Let $u = b$. Then the neighbors of all highlighted vertices are $F = \{a, c, d, e\}$. Note, we only consider updating the labels for a and c since these vertices are adjacent to b, the vertex highlighted in the previous step.

$$w(b) + w(ba) = 3 + 6 = 9 \not< 9 = w(a)$$
$$w(b) + w(bc) = 3 + 5 = 8 < \infty = w(c)$$

We do not update the label for a, but the label for c is updated to $L(c) = (b, 8)$. Since the minimum weight for all vertices in F is that of d, we highlight the edge fd and the vertex d.

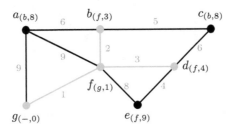

Step 5: Let $u = d$. Then the neighbors of all highlighted vertices are $F = \{a, c, e\}$. We compute

$$w(d) + w(dc) = 4 + 6 = 10 \not< 7 = w(c)$$
$$w(d) + w(de) = 4 + 4 = 8 < 9 = w(e)$$

Update $L(e) = (d, 8)$. Since the minimum weight for all vertices in F is that of c, we highlight the edge bc and the vertex c. This terminates the iterations of the algorithm since our ending vertex has been reached.

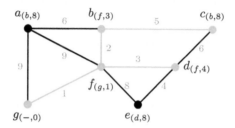

Output: The shortest path from g to c is $g\,f\,bc$, shown in blue above. This path has a total weight 7, as shown by the label of c.

Although the form given above for Dijkstra's Algorithm is written for an undirected graph, with very little modification it can be applied to a digraph as well. Recall from Section 2.3, that a digraph is a graph in which the edges now have a direction associated to them, which could be used to model a one-way street. If you have not already done so, you are encouraged to read back through Section 2.3. The example below shows how to apply Dijkstra's Algorithm to a digraph, where the main change is that the neighbor set F only consists of vertices with an arc from a previously highlighted vertex.

Example 3.3 Apply Dijkstra's Algorithm to the digraph below where $Start = g$ and $End = c$.

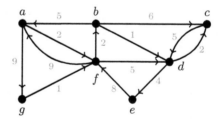

Solution:

Step 1: Highlight g. Define $L(g) = (-, 0)$ and $L(x) = (-, \infty)$ for all $x = a, \cdots, f$.

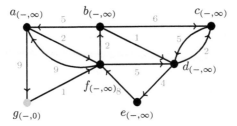

Step 2: Let $u = g$. Then the neighbors of g are $F = \{f\}$. We compute

$$w(g) + w(gf) = 0 + 1 = 1 < \infty = w(f)$$

Update $L(f) = (g, 1)$ and highlight the arc gf and the vertex f.

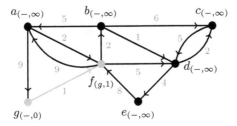

Step 3: Let $u = f$. Then the neighbors of all highlighted vertices are $F = \{a, b, d\}$. We compute

$$w(f) + w(fa) = 1 + 9 = 10 < \infty = w(a)$$
$$w(f) + w(fb) = 1 + 2 = 3 < \infty = w(b)$$
$$w(f) + w(fd) = 1 + 5 = 6 < \infty = w(d)$$

Update $L(a) = (f, 10)$, $L(b) = (f, 3)$, and $L(d) = (f, 6)$. Since the minimum weight for all vertices in F is that of b, we highlight the arc fb and the vertex b.

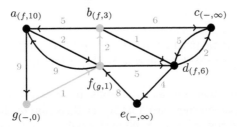

Step 4: Let $u = b$. Then the neighbors of all highlighted vertices are $F = \{a, c, d\}$. We compute

$$w(b) + w(ba) = 3 + 5 = 8 < 10 = w(a)$$
$$w(b) + w(bc) = 3 + 6 = 9 < \infty = w(c)$$
$$w(b) + w(bd) = 3 + 1 = 4 < 6 = w(d)$$

Update $L(a) = (b, 8)$, $L(c) = (b, 9)$, and $L(d) = (b, 4)$. Since the minimum weight for all vertices in F is that of d, we highlight the arc bd and the vertex d.

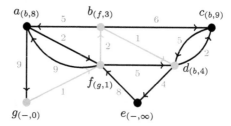

Step 5: Let $u = d$. Then the neighbors of all highlighted vertices are $F = \{a, c, e\}$. We compute

$$w(d) + w(dc) = 4 + 2 = 6 < 9 = w(c)$$
$$w(d) + w(de) = 4 + 4 = 8 < \infty = w(e)$$

Update $w(c) = (d, 6)$ and $L(e) = (d, 8)$. Since the minimum weight for all vertices in F is that of c, we highlight the edge dc and the vertex c. This concludes the algorithm since the ending vertex has been reached.

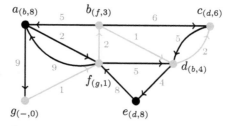

Output: The shortest path from g to c is $g\,f\,b\,d\,c$, shown in blue above. This path has a total weight 6, as shown by the weight of c.

One final note about paths in digraphs. It is possible for a path not to exist from one vertex to another based upon the direction of the arcs (for example, if all arcs pointed toward a, then no path originating at a could exist). In such a situation, Dijkstra's Algorithm would halt and note that a shortest path could not be found.

Chinese Postman Problem Revisited

Section 1.5 discussed the Chinese Postman Problem, which consisted of finding an exhaustive circuit of minimal total weight through a weighted graph. The initial steps were determining which vertices had odd degree and pairing these in the hopes of minimizing the weight along the path between each pair. The examples studied in Chapter 1 did not need the complexity of Dijkstra's Algorithm in finding the shortest paths. Below we combine the method for Eulerizing a graph and Dijkstra's Algorithm to provide a more complete answer to the Chinese Postman Problem.

Example 3.4 The graph below represents a town in which a postman must deliver the mail, and so he must travel each edge at least once. Use Dijkstra's Algorithm to find the best pairing of odd vertices and the total weight of the edges duplicated in the Eulerization of the graph.

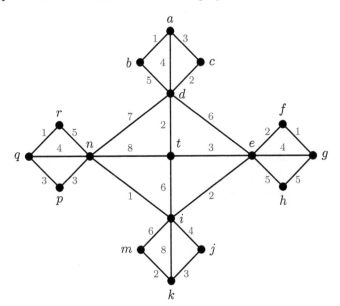

Solution: There are four odd vertices that must be paired in the optimal Eulerization, namely $a, g, k,$ and q. Three possible pairings of these vertices exist: $a - g$ and $k - q$, $a - k$ and $g - q$, $a - q$ and $g - k$. Applying Dijkstra's Algorithm, we can find the shortest paths between the paired vertices and the total weight of the two paths needed to Eulerize the graph (the details are left as Exercise 3.3).

Path Pairs	Weight	Total Weight
$a\,d\,t\,e\,f\,g$	12	
$k\,j\,i\,n\,q$	12	24
$a\,d\,t\,e\,i\,j\,k$	18	
$g\,f\,e\,i\,n\,q$	10	28
$a\,d\,n\,q$	15	
$g\,f\,e\,i\,j\,k$	12	27

Once the paths are found, we duplicate the edges along these paths to obtain the Eulerization. In doing so, we must be on alert for any edges that appear in both paths of a pairing. For example, the paths in the second pairing both use the edge $e\,i$, as shown on the graph below on the left where the path from a to k is given by the edges with one arrow and the path from g to q the edges with two arrows. As noted in Chapter 1, we should never duplicate an edge more than once during an Eulerization. We modify the paths found by Dijkstra's Algorithm by removing both duplications of $e\,i$. This maintains the degree condition (all vertices have even degree) and reduces the total weight by 4. The second pairing now results in the same Eulerization as that of the first pairing and has a total weight increase of 24, as shown on the graph below on the right. This provides the optimal Eulerization as required.

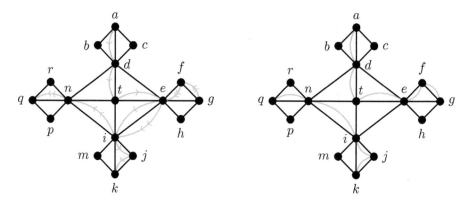

The use of Dijkstra's Algorithm allows for a methodical approach to the Chinese Postman Problem. However, even for small examples the number of paths to find can be quite large. In fact, the number of possible ways to pair n vertices of odd degree (where n is even) is $(n-1)!!$, called n *double factorial*. For a given integer k, $k!!$ is defined as the product of all even integers less than or equal to k if k is even and the product of all odd integers less than or equal to k if k is odd. Once the groups of pairings are formed, Dijkstra's Algorithm will be applied $n/2$ times to find the shortest paths between paired vertices. In Example 3.4 above, four vertices required $3!! = 3 \cdot 1 = 3$ possible groups of pairings, each of which contained two pairs, and so 6 possible paths needed to be calculated. If there were 8 odd vertices, then we would calculate $\frac{8}{2} \cdot 7!! = 4 \cdot 7 \cdot 5 \cdot 3 \cdot 1 = 420$ paths. This approach is inefficient and grows quite quickly (along the magnitude of Brute Force from Chapter 2). The optimal approach, which can be seen in [13], uses shortest paths but with a matching algorithm (matchings will be discussed in Chapter 5) that reduces the complexity and results in an efficient algorithm.

3.2 Project Scheduling

This section still focuses on finding a specific path within a graph, but we are no longer interested in the shortest path. Instead, we will be searching for a *critical path* within a graph representing multiple interdependent pieces of a project. Critical paths will be defined later, but for now consider the following scenario:

> You are hosting a back to school party and a few friends have offered to help with the preparations. You need to buy and cook the food, buy and put out the drinks (and you prefer to set up the ice buckets once the food is cooked so the drinks stay cold), dust and vacuum the house (and you always vacuum after the dusting), and set the table (which must be completed after the vacuuming and cooking is done). What is the best way to finish the preparations on time and with as little help as possible?

The project above contains multiple pieces that have varying levels of interdependency and we must determine the best way to assign the various tasks to the people available. To visualize the relationship between tasks in a scheduling problem, we will use a digraph where the vertices represent the individual tasks and an arc from a to b indicates that task a must be completed before b can begin. Note that although the example above is grounded in the real world, project scheduling is primarily used in computer programming and the terminology reflects this.

Definition 3.1 Consider a project containing multiple parts or steps.

- A *task* in a required step of a project that cannot be broken into smaller pieces. These will be labeled with lowercase letters (e.g., a, b, c).

- A *processor* is the unit (such as a person) that completes a task. Processors will be labeled as P_1, P_2, P_3, etc. At any time a processor will either be idle or busy performing a task.

- At any stage of a project, a task can be in one of four states: *eligible*, *ineligible*, *in execution*, or *completed*. A task is eligible when all the tasks it relies upon are completed.

- The *processing time* of a task is the time it takes to complete the task, denoted by $pt(v)$ for task v.

- If task b relies on the completion of task a before it can be eligible, we call this a *precedence relationship*.

- The *finishing time* of a schedule is the total time used in that schedule. The *optimal time* of a project is the minimum finishing time among all possible schedules, denoted OPT.

In Example 3.5 below, the project introduced above is reworded to make use of this new terminology. In addition, the information is displayed in digraph form and a method for assigning tasks to processors is discussed in Example 3.6.

Example 3.5 Look back at the example on party planning described at the start of this section. The table below provides the precedence relationships and processing times of the tasks. The processing time is given in minutes. Model the information with a digraph.

Task	Vertex Name	Processing Time	Precedence Relationships
Buy Food	f	40	
Buy Drinks	b	20	
Dust	d	20	
Vacuum	v	30	d
Cook Food	c	60	f
Set Out Drinks	s	30	b, c
Set Table	t	20	v, c

Solution: It is customary to include a vertex to represent the start and end of a project, as well as lay out vertices to avoid edge crossings whenever possible. The processing times are shown in parentheses next to the vertex labels.

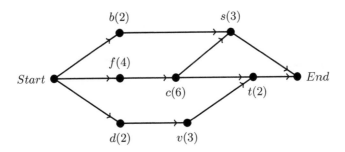

Once a digraph has been created, the next step is to determine which processor (or person) should complete each task. This may be easy in a project with only a few tasks, or if the interplay between tasks is not complex. However, as complexity grows, we will need a procedure for assigning tasks to people. We will use a **Priority List Model** for scheduling, which consists of establishing an ordering of the tasks into a list. Tasks must be assigned to processors according to their order in the priority list while precedence relationships, which are displayed in the digraph, are used to determine eligibility of a given task. Later, we will discuss good approaches to finding a priority list.

Example 3.6 Using the priority list $b - d - t - v - s - f - c$, find a schedule for the project from Example 3.5 using two processors.

Solution: Each step represents a moment in time where a decision must be made. The time in question will be noted in parentheses at the start of each step.

Step 1: (T=0) The first item in the priority list is b. Since b does not rely on any other task, assign it to P_1. The next item in the priority list, d, is also eligible. Assign d to P_2.

	10	20	30	40	50	60	70	80	90	100	110	120	130	140	150
P_1	b	b													
P_2	d	d													

Step 2: (T=20) The next point at which a processor is free to pick up a task is at 20 minutes. Search the priority list to find the next eligible task. Although t is the next item in the list, it is ineligible since neither c nor v is complete. Since d is complete, v is eligible. Assign v to P_1. Since P_2 is also ready for a new task, search again for the next eligible task in the list, which is f, and assign it to P_2.

	10	20	30	40	50	60	70	80	90	100	110	120	130	140	150
P_1	b	b	v	v	v										
P_2	d	d	f	f	f	f									

Step 3: (T=60) At 60 minutes, Processor 1 is ready for a new task. Search through the priority list for the next eligible task. Unfortunately, all other tasks require f to be complete before they can begin. P_1 will remain idle until f is completed. Idle time will be noted by an asterisk ($*$).

	10	20	30	40	50	60	70	80	90	100	110	120	130	140	150
P_1	b	b	v	v	v	*									
P_2	d	d	f	f	f	f									

Step 4: (T=70) At 70 minutes, both processors are ready to take up a new task. The only eligible item is c. When more than one processor is ready to start a new task but only one task is available, by convention we assign the task to the lower indexed processor. The other processor remains idle.

	10	20	30	40	50	60	70	80	90	100	110	120	130	140	150
P_1	b	b	v	v	v	*	c	c	c	c	c	c			
P_2	d	d	f	f	f	f	*	*	*	*	*	*			

Step 5: (T=130) Once task c is complete, we can assign task t to P_1 and task s to P_2.

	10	20	30	40	50	60	70	80	90	100	110	120	130	140	150
P_1	b	b	v	v	v	*	c	c	c	c	c	c	t	t	*
P_2	d	d	f	f	f	f	*	*	*	*	*	*	s	s	s

The priority list $b - d - t - v - s - f - c$ yields a finishing time of 150 minutes using two processors.

A few items should stand out in the example above. First, although item t came before v in the priority list we could not place it into the schedule at time 20 since it was ineligible until both c and v were complete. Hence, we skipped over t and moved to the next eligible item in the priority list. Next, the schedule we obtained contains a large amount of idle time, 8 hours in total. Although some idle time may be unavoidable, its presence should indicate that more investigation is warranted. Finally, the priority list given did not seem to have any connection with the digraph (in fact, it was generated randomly). A better approach would be to use information from the digraph to obtain a good priority list. To do that, we make use of a specific directed path within a digraph.

Critical Paths

The first part of the chapter focused on finding the shortest path within a graph (or digraph). For project scheduling, we are looking for the opposite — the longest paths within the project digraph. For example, the path $Start \rightarrow d \rightarrow v \rightarrow t \rightarrow End$ from the digraph in Example 3.5 has a total time of 70 minutes and so the party preparations cannot be completed in less time. Rather than looking at all possible paths, we focus on finding a ***critical path***. The critical path is the path with the highest total time out of all paths that begin at vertex *Start* and finish at *End*. This path is of interest because it easily identifies restrictions on the completion time of a project. In addition, it indicates which tasks should be prioritized. To find the critical path, we first need to find the *critical times* of all vertices in the graph.

Definition 3.2 The ***critical time*** $ct[x]$ of a vertex x is defined as the sum of the processing time of x with the maximum of the critical times for all vertices y for which xy is an arc. That is,

$$ct[x] = pt(x) + maximum\{ct[y], \text{ for all } y \text{ where } xy \text{ is an arc}\}.$$

Once the critical times are obtained, a critical path is found and the critical path priority list is created. The algorithm below outlines the process of finding the critical times, a critical path, and the priority list.

Critical Path Algorithm

Input: Project digraph G with processing times given.

Steps:

1. Label the vertex *End* with $pt(0)$ and $ct[0]$. For any vertex x with an arc to *End*, define $ct[x] = pt(x)$.

2. Travel the arcs in reverse order. When a new vertex is encountered, calculate its critical time.

3. Once all critical times have been obtained, find the path from *Start* to *End* where if more than one arc exits out of a vertex, take the arc to the neighbor vertex of largest critical time.

4. Create a priority list by ordering vertices by decreasing critical time.

Output: Critical path and critical path priority list.

Example 3.7 Apply the Critical Path Algorithm to the project digraph from Example 3.5.

Solution: It is customary to use brackets for the critical times, distinguishing them from the processing times.

Step 1: Label *End* with critical time 0. Since s and t have arcs to End, set $ct[s] = pt(s) = 3$ and $ct[t] = pt(t) = 2$.

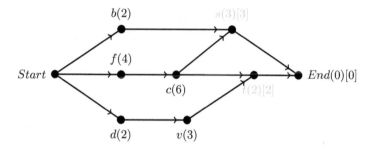

Step 2: As v has a single arc to t, define the critical time of v as

$$ct[v] = pt(v) + ct[t] = 3 + 2 = 5.$$

The vertex c has an arc to both s and t. Since $ct[s] > ct[t]$ we get

$$ct[c] = pt(c) + ct[s] = 6 + 3 = 9.$$

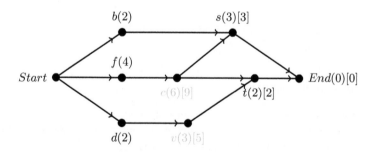

Step 3: The remaining three vertices each have a single arc to previously considered vertices. Define

$$ct[b] = pt(b) + ct[s] = 2 + 3 = 5$$
$$ct[f] = pt(f) + ct[c] = 4 + 9 = 13$$
$$ct[d] = pt(d) + ct[v] = 2 + 5 = 7$$

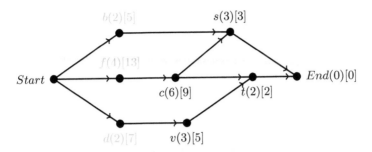

Step 4: Label the processing time of *Start* as 0 and the critical time as 13 since f is the neighbor with the largest critical time.

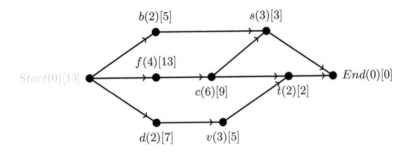

Step 5: Follow the path from *Start* to *End* where the vertices are chosen based on the largest critical times. This gives the path $Start \to f \to c \to s \to End$ of total time 130, which is highlighted below. Ordering the vertices in decreasing order of critical times gives the critical path priority list of $f-c-d-b-v-s-t$.

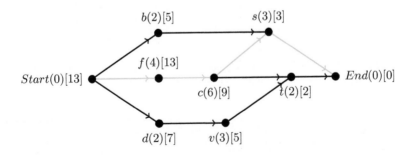

Now that the critical path priority list is complete, we can find a schedule for the project. This will provide a better schedule than the random one from Example 3.6. In general, the critical path priority list results in a very good, though not always optimal, schedule.

Example 3.8 Use the critical path priority list from Example 3.7 to find a schedule using two processors.

Solution: The critical path priority list is $f - c - d - b - v - s - t$.

Step 1: (T=0) Since f is the first item in the list, assign it to P_1. We cannot assign c to P_2 since it relies on the completion of f. Moving to the next eligible task puts d into P_2.

	10	20	30	40	50	60	70	80	90	100	110	120	130	140	150
P_1	f	f	f	f											
P_2	d	d													

Step 2: (T=20) P_2 has completed d and can be assigned a new task. Since f is still not complete, assign b, the next eligible task, to P_2.

	10	20	30	40	50	60	70	80	90	100	110	120	130	140	150
P_1	f	f	f	f											
P_2	d	d	b	b											

Step 3: (T=40) Both P_1 and P_2 can be assigned a new task. Task c is now eligible and will be assigned to P_1. The next eligible task is v; assign it to P_2.

	10	20	30	40	50	60	70	80	90	100	110	120	130	140	150
P_1	f	f	f	f	c	c	c	c	c	c					
P_2	d	d	b	b	v	v	v								

Step 4: (T=70) P_2 can be assigned a new task. However, all remaining tasks are ineligible since they rely on the completion of c. P_2 remains idle.

	10	20	30	40	50	60	70	80	90	100	110	120	130	140	150
P_1	f	f	f	f	c	c	c	c	c	c					
P_2	d	d	b	b	v	v	v	*	*	*					

Step 5: (T=100) Both P_1 and P_2 can be assigned a new task. Since s is now eligible and first in the priority list, it is assigned to P_1. Task t is also eligible and assigned to P_2.

	10	20	30	40	50	60	70	80	90	100	110	120	130	140	150
P_1	f	f	f	f	c	c	c	c	c	c	s	s	s		
P_2	d	d	b	b	v	v	v	*	*	*	t	t	*		

The schedule above has a finishing time of 130 minutes and contains 4 total hours of idle time.

Compare the schedule obtained using the Critical Path Algorithm with the one using a random priority list from Example 3.6. A few observations should stand out. Both schedules contained some idle time, though the one utilizing the critical path priority list had half that of the initial example. This is in part because items on the critical path were prioritized over less important tasks. Moreover, the schedule above must be optimal since its finishing time is equal to the critical time of *Start*.

In general, it can be difficult to determine if a schedule is optimal (you could use brute force and find all possible priority lists and their resulting schedules but there are $n!$ possible lists for a project with n items). However, we have two very quick calculations that can help identify when a schedule is optimal:

- The optimal time of a schedule is no less than the critical time of *Start*; that is, $OPT \geq cr[Start]$.

- Calculate the sum of all processing times of all tasks. The optimal time is no less than this sum divided by the total number of processors used; that is, $OPT \geq \dfrac{\text{sum of processing times}}{\text{number of processors}}$.

The first item above is not impacted by how many processors are being used, whereas the second calculation may change depending on the number of processors available. For example, in the party preparations the sum of all processing times is 220 minutes. Using two processors gives $OPT \geq \frac{220}{2} = 110$ and using three processors gives $OPT \geq \frac{220}{3} \approx 73$. However, since $cr[Start] = 130$, neither of these calculations provides additional insight into the optimal schedule.

When determining if a schedule is optimal, there are two questions you must ask: is the number of processors fixed or can the number of processors be increased to give a shorter finishing time? What is more important, finishing in the shortest time possible or reducing the number of processors needed? This book does not provide answers to these questions, but rather the mathematical tools for supporting your answer.

Example 3.9 This spring you are tackling the jungle that is your backyard. Some good friends have volunteered their time and you have split them into two groups. You will need to buy plants, remove the old bushes and ivy along the back fence, weed the flower beds, plant and fertilize the new bushes, plant flowers, trim trees, mow and rake the lawn, and install solar powered path lighting. The table below lists these tasks, their expected time, and any precedence relationship that exists between tasks. Model this using a digraph, apply the Critical Path Algorithm and construct a schedule. Determine if your schedule is optimal.

	Vertex	Processing	Precedence
Task	Name	Time	Relationships
Buy plants	b	1	
Remove bushes	r	7	
Remove ivy	i	4	
Weed flower beds	w	3	b, r
Plant bushes	p	7	b, r
Plant flowers	f	1	w, p
Trim trees	t	4	i
Mow and rake lawn	m	6	t
Install lighting	l	2	w

Solution: The digraph is given below with the processing times in parentheses and the critical times in brackets.

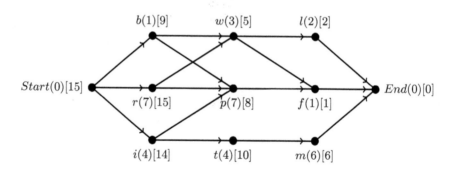

From the digraph, we get the critical path

$$Start \rightarrow r \rightarrow p \rightarrow f \rightarrow End.$$

The critical path priority list is $r - i - t - b - p - m - w - l - f$ and gives the schedule shown below.

	1	2	3	4	5	6	7	8	9	10	11	12	13	14	15	16	17	18	19
P_1	r	r	r	r	r	r	r	b	p	p	p	p	p	p	p	*	*	l	l
P_2	i	i	i	i	t	t	t	t	m	m	m	m	m	m	w	w	w	f	*

By using the two metrics described above, we know $OPT \geq 15$ and $OPT \geq 17.5$ when using two processors. Since there is no half-unit of time, we know the optimal time in fact must be at least 18 hours. The schedule above has 3 total hours of idle time. The optimal schedule is shown below.

	1	2	3	4	5	6	7	8	9	10	11	12	13	14	15	16	17	18
P_1	r	r	r	r	r	r	r	p	p	p	p	p	p	p	l	l	f	*
P_2	i	i	i	i	b	t	t	t	t	w	w	w	m	m	m	m	m	m

Note that w did not place high on the priority list, but needed to be completed earlier since both l and f relied upon its completion. By making this small modification, we were able to find the optimal schedule using two processors. Exercise 3.6 asks for a schedule using three processors and Exercise 3.7 asks you to verify the critical times.

3.3 Exercises

3.1 Apply Dijkstra's Algorithm to each of the graphs below, where the starting vertex is x and the ending vertex is y.

(a)   (b)

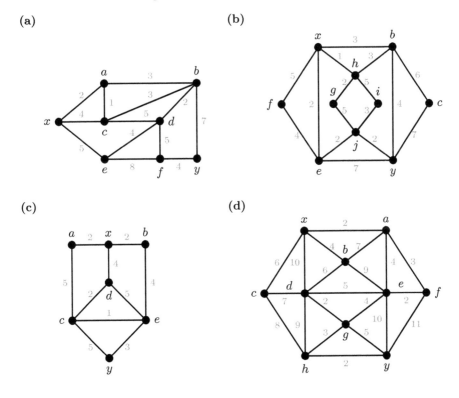

(c)   (d)

(e)

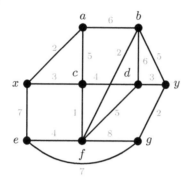

(f)

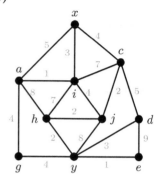

(g)

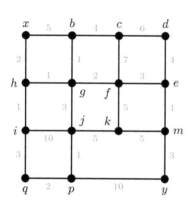

3.2 Modify Dijkstra's Algorithm so that the output is the shortest path from the desired vertex to all other vertices in the graph.

3.3 Apply Dijkstra's Algorithm to the graph from Example 3.4 and verify the shortest paths are listed in the solution.

3.4 Apply Dijkstra's Algorithm to each of the digraphs below, where the starting vertex is x and the ending vertex is y.

(a)

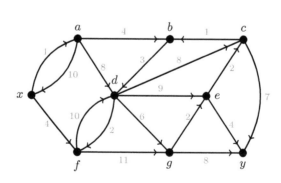

(b)

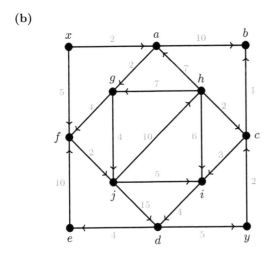

(c)

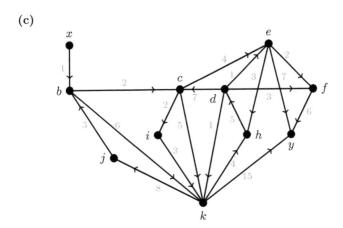

3.5 Explain why a third processor for Example 3.5 would be unnecessary.

3.6 Find a schedule using three processors for the project outlined in Example 3.9. Compare your answer to the schedule provided using two processors.

3.7 Verify the critical times for Example 3.9.

3.8 Consider the project digraph shown below.

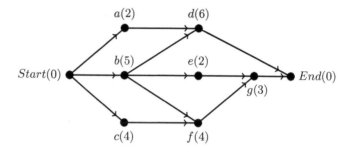

(a) Use the Critical Path Algorithm to find a schedule with 2 processors.
(b) Determine if the schedule is optimal. If not, find a better schedule using 2 processors.

3.9 Consider the project digraph shown below.

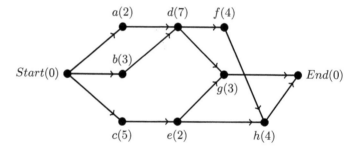

(a) Use the Critical Path Algorithm to find a schedule with 2 processors.
(b) Determine if the schedule is optimal. If not, find a better schedule using 2 processors.

3.10 Consider the project digraph shown below.

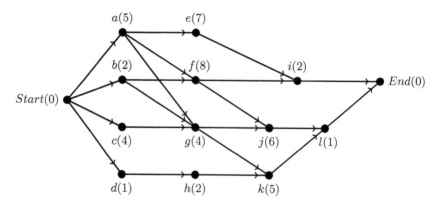

(a) Use the Critical Path Algorithm to find a schedule with 2 processors.
(b) Use the Critical Path Algorithm to find a schedule with 3 processors.
(c) Determine if either schedule is optimal.

3.11 The table below lists 9 tasks that comprise a project, as well as their processing times and precedence relationships.

Task	Processing Time	Precedence Relationships
a	2	
b	5	
c	1	a
d	2	a, b
e	4	b
f	6	c, e
g	7	d
h	6	e
i	2	f, g

(a) Draw the project digraph.
(b) Use the Critical Path Algorithm to find a schedule with 2 processors.
(c) Use the Critical Path Algorithm to find a schedule with 3 processors.
(d) Determine if either schedule is optimal.

3.12 The table below lists 10 tasks that comprise a project, as well as their processing times and precedence relationships.

Task	Processing Time	Precedence Relationships
a	2	
b	4	
c	6	a
d	5	a
e	4	a, b
f	5	b
g	2	b
h	10	c
i	3	d, e, f
j	4	f, g

(a) Draw the project digraph.
(b) Use the Critical Path Algorithm to find a schedule with 2 processors.

(c) Use the Critical Path Algorithm to find a schedule with 3 processors.
(d) Determine if either schedule is optimal.

3.13 The table below lists 13 tasks that comprise a project, as well as their processing times and precedence relationships.

Task	Processing Time	Precedence Relationships
a	2	
b	3	
c	1	
d	3	
e	4	a
f	6	b
g	4	c, d
h	4	e
i	3	e, f, g
j	2	e, f, g
k	11	d
l	2	h
m	1	i, j

(a) Draw the project digraph.
(b) Use the Critical Path Algorithm to find a schedule with 2 processors.
(c) Use the Critical Path Algorithm to find a schedule with 3 processors.
(d) Determine if either schedule is optimal.

Projects

3.14 You are starting a new business that will serve customers in 7 different cities. You need to determine the distance from your hub location to each of your customers. Choose 7 cities from the table at the end of Chapter 2 (on pages 79 and 80) and choose one to be your hub location. Use the modification of Dijkstra's Algorithm from Exercise 3.2 to minimize the mileage from your hub location to each of the customer locations.

3.15 You have decided to start a small business that manufactures a product. You are trying to determine the number of employees you will need, as well as how long it will take to produce one item. Write a report where Critical Paths are used to answer these questions. To complete the report, you will need to name your business, the product, determine the different stages of production and their precedence relationships, and draw the project digraph. Your product should have at least 10 different tasks and at least 2 tasks must rely on multiple tasks. Form schedules using the Critical Path Algorithm using 2, 3, and 4 processors.

Chapter 4

Trees and Networks

The three previous chapters focused on finding routes within a graph, whether they were exhaustive (such as visiting every edge or every vertex), optimal (shortest paths), or of greatest importance (critical paths). Consider the following scenario:

> Jim's start-up company recently secured investors, allowing him to hire more employees and retrofit the new office space. The cost of setting up and maintaining a computer network is directly related to the amount of cable installed, so Jim wishes to use as little cable as possible while still ensuring all employees can access the company's central server.

It should be clear that this is *not* a routing problem; we are not interested in visiting a sequence of edges or vertices, but rather ensure that all locations are connected. What graph model solves this problem? As we will see in this chapter, Jim needs to find a specific type of graph called a tree.

This seemingly different graph model and departure from previous routing problems has strong connections to paths from Chapter 3, as well as a strategy for solving an instance of the Traveling Salesman Problem from Chapter 2.

4.1 Trees

The scenario above is only concerned with ensuring the graph is connected, or to put it another way, there must be a path between any two vertices. You could obviously create a complete graph on the vertices, which would translate to connecting every computer to all possible ports. Though this guarantees every computer can access the central server, more cable than necessary would be used. In effect, we want a graph where each computer has a single path back to the central server, or in graph theoretic terms we do not want any cycles. These graphs are called *trees*.

Definition 4.1 A graph G is

- *acyclic* if there are no cycles or circuits in the graph.

- a *network* if it is connected.

- a *tree* if it is an acyclic network; that is, a graph that is both acyclic and connected.

- a *forest* if it is an acyclic graph.

In addition, a vertex of degree 1 is called a *leaf.*

Trees arise in many seemingly unrelated disciplines, including probability, chemistry, and computer science. The next few examples provide context for the interest in trees, but are in no way comprehensive in terms of the applications of trees. After which, we will discuss optimal trees and how to find them.

Example 4.1 Adam comes to you with a new game. He flips a coin and you roll a die. If he gets heads and you roll an even number, you win \$2; if he gets heads and you roll an odd number, you pay him \$3. If he gets tails and you roll either 1 or 4, you win \$5; if he gets tails and any of 2, 3, 5, or 6 is rolled, you pay him \$2. What is the probability you win \$5? What is the probability you win any amount of money?

Solution:
A probability tree consists of vertices representing the possible outcomes of each part of the experiment (here a coin and dice game) and the edges are labeled with the probability that the outcome occurred. To find the probability of any final outcome, multiply along the path from the initial vertex to the ending result. The tree below has the edges labeled and the final probabilities calculated for Adam's game.

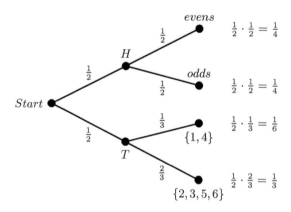

Using the tree above, the probability you win \$5 (which requires tails and a 1 or 4) is $\frac{1}{6}$ and the probability you win any money is $\frac{1}{6} + \frac{1}{4} = \frac{5}{12}$. Do not play this game with Adam! He is more likely to win than you!

Example 4.2 *Chemical Graph Theory* uses concepts from graph theory to obtain results about chemical compounds. In particular, individual atoms in a molecule are represented by vertices and an edge denotes a bond between the atoms. One way to determine the number of isomers for a molecule is to determine the number of distinct graphs that contain the correct type of each atom. For hydrocarbons (molecules only containing carbon and hydrogen atoms) the *hydrogen-depleted* graph is used since the bonds between the carbon atoms will uniquely determine the locations of the hydrogen atoms.

Below are the only two trees on four vertices. These correspond to the only possible isomers of butane (C_4H_{10}), namely n-butane ($H_3C(CH_2)_2CH_3$) and isobutane ($(H_3C)_3CH$), whose full molecular forms are displayed below their respective hydrogen-depleted graph. By using graph theory, we can prove no other isomers of butane are possible since no other trees on four vertices exist. [1]

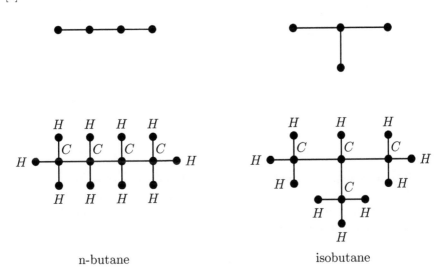

n-butane   isobutane

Example 4.3 Trees can be used to store information for quick access. Consider the following string of numbers:

$$4, 2, 7, 10, 1, 3, 5$$

We can form a tree by creating a vertex for each number in the list. As we move from one entry in the list to the next, we place an item below and to the left if it is less than the previously viewed vertex and below and to the right if it is greater. If we add the restriction that no vertex can have more than two edges coming down from it, then we are forming a binary tree.

For the string above, we start with a tree consisting of one vertex, labeled 4 (see T_1 below). The next item in the list is a 2, which is less than 4 and so its vertex is placed on the left and below the vertex for 4. The next item, 7, is larger than 4 and so its vertex is placed on the right and below the vertex for 4 (see T_2 below).

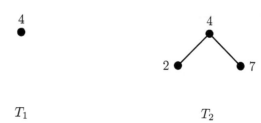

T_1 $\qquad\qquad\qquad\qquad$ T_2

The next item in the list is 10. Since 4 already has two edges below it, we must attach the vertex for 10 to either 2 or 7. Since 10 is greater than 4, it must be placed to the right of 4 and since 10 is greater than 7, it must be placed to the right of 7 (see T_3 below). A similar reasoning places 1 to the left and below 2 (see T_4 below).

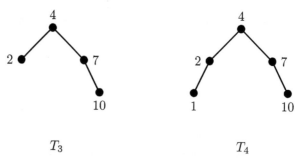

T_3 $\qquad\qquad\qquad\qquad$ T_4

The next item is 3, which is less than 4 and so must be to the left of 4. Since 3 is greater than 2, it must be placed to the right of 2 (see T_5 below). The final item is 5, which is greater than 4 but less than 7, placing it to the right of 4 but to the left of 7 (see T_6 below).

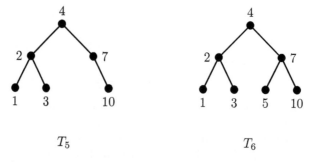

T_5 $\qquad\qquad\qquad\qquad$ T_6

This final tree T_6 represents the items in our list. If we want to search for an item, then we only need to make comparisons with at most half of the items in the list. For example, if we want to find item 5, we first compare it to the vertex at the top of the tree. Since 5 is greater than 4, we move along the edge to the right of 4 and now compare 5 to this new vertex. Since 5 is less than 7 we move along the edge to the left of 7 and reach the item of interest. This allows us to find 5 by making only two comparisons rather than searching through the entire list.

This searching technique can be thought of as searching for a word in a dictionary. Once you open to a page close to the word you are searching for, you flip pages back and forth depending on if you are before or after the word needed. Searching through the list item-by-item would be like flipping the pages one at a time, starting from the beginning of the dictionary, until you find the correct word (not a very efficient method).

The tree obtained in the example above is referred to as a *rooted tree*. Rooted trees are often used when there is a clear starting point for the tree, such as in a decision tree or family tree. For additional information on rooted trees, see Section 7.5.

Due to their specialized nature, trees contain many unique properties, similar to those of complete graphs from Chapter 2.

Properties of Trees

(1) For every $n \geq 1$, any tree with n vertices has $n - 1$ edges.

(2) For any tree with $n \geq 1$ vertices, the sum of the degrees is $2n - 2$.

(3) Every tree with at least two vertices contains at least two leaves.

(4) Any network on n vertices with $n - 1$ edges must be a tree.

(5) For any two vertices in a tree, there is a unique path between them.

(6) The removal of any edge of a tree will disconnect the graph.

These properties can be especially helpful when using algorithms to find a tree within a larger graph. In particular, given a set number of vertices we know the correct number of edges needed in a tree as well as the proper degrees within a tree. Further uses of these properties appear in the exercises.

4.2 Spanning Trees

Returning to the scenario first proposed at the start of this chapter, Jim's predicament can be thought of as building a tree for his new office space.

However, since every possible connection is available, he must choose those with the least length. Jim wants to find an underlying tree structure from a weighted complete graph. These are called *spanning trees*.

Definition 4.2 A *subgraph* H of a graph G is a graph where H contains some of the edges and vertices of G; that is, $V(H) \subseteq V(G)$ and $E(H) \subseteq E(G)$.

We say H is a *spanning subgraph* if it contains all the vertices but not necessarily all the edges of G; that is, $V(H) = V(G)$ and $E(H) \subseteq E(G)$.

A *spanning tree* is a spanning subgraph that is also a tree.

Note that if an edge appears in a subgraph, then both endpoints must also be included in the subgraph. However, if a vertex appears in a subgraph, any number of its incident edges may be included.

Example 4.4 For each of the graphs below, find a spanning tree and a subgraph that does not span.

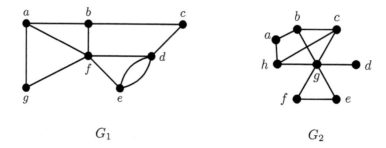

$$G_1 \qquad\qquad\qquad G_2$$

Solution: To find a spanning tree, we must form a subgraph that is connected, acyclic, and includes every vertex from the original graph. The graphs T_1 and T_2 below are two examples of spanning trees for their respective graphs; other examples exist.

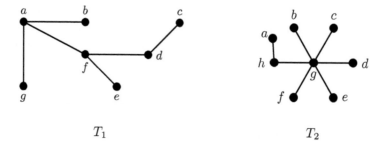

$$T_1 \qquad\qquad\qquad T_2$$

The subgraph H_1 below is neither spanning nor a tree since some vertices from G_1 are missing and there is a multi-edge (and hence a circuit) between d and e. The subgraph H_2 below is not spanning since it does not contain vertex a but is a tree since no circuits or cycles exist. As above, these are merely examples and other non-spanning subgraphs exist.

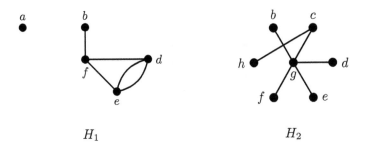

H_1 $\qquad\qquad\qquad\qquad\qquad$ H_2

Kruskal's Algorithm

Similar to Dijkstra's Algorithm studied in Chapter 3, Kruskal's Algorithm is fairly modern, first published in 1956. Joseph Kruskal was an American mathematician best known for his work in statistics and computer science. This algorithm is unique in that it is both efficient and optimal while still easily implemented and understandable for a non-scientist. In fact, it is the preferred method for finding a minimum spanning tree when the edges can be easily sorted.[19]

Kruskal's Algorithm

Input: Weighted connected graph $G = (V, E)$.

Steps:

1. Choose the edge of least weight. Highlight it and add it to $T = (V, E')$.

2. Repeat step (1) so long as no circuit is created. That is, keep picking the edges of least weight but skip over any that would create a cycle in T.

Output: Minimum spanning tree T of G.

Kruskal's algorithm does not distinguish between two edges of the same weight, in part because it does not influence the outcome. If at any point there are two edges to choose from of the same weight, you can pick either one. In addition, at each step of the algorithm we are building a forest subgraph that will eventually result in a spanning tree.

Example 4.5 Find the minimum spanning tree of the graph G below using Kruskal's Algorithm.

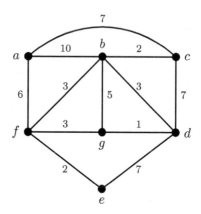

Solution: At each step, the newest edge added will be highlighted in blue and the previously chosen edges will be in black. Unchosen edges will be shown in gray.

Step 1: Pick the smallest edge, gd and highlight it.

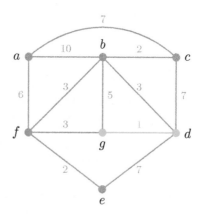

Step 2: Pick the next smallest edge. There are two edges of weight 2. Either is a valid choice. We choose bc.

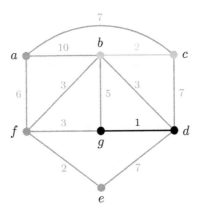

Step 3: The other edge of weight 2, ef, is still a valid choice. Add it to T.

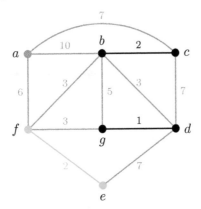

Step 4: The next smallest edge weight is 3, and there are 3 edges to choose from. We randomly pick bf.

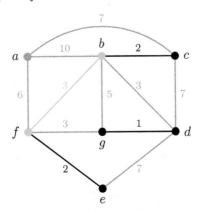

Step 5: Both of the other edges of weight 3 are still available. We choose bd.

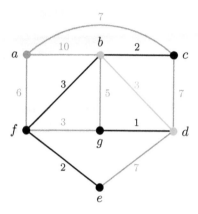

Step 6: At this point, we cannot choose the last edge of weight 3, fg, since it would create a circuit (namely, $b\,d\,g\,f\,b$).

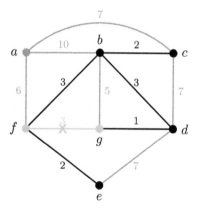

The next smallest edge is bg of weight 5. Again, we cannot choose this edge since it would create a circuit ($b\,d\,g\,b$).

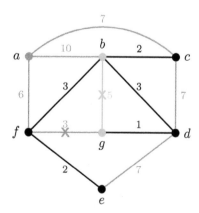

The next available edge is af of weight 6. This is also the last edge needed since we now have a tree containing all the vertices of G. In addition, we know this is a tree since we have 7 vertices and 6 edges.

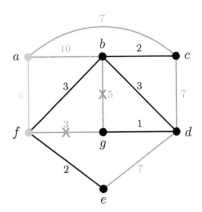

Output: The tree below with total weight 17.

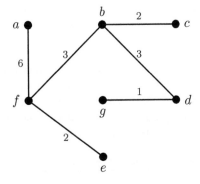

In Steps 4 and 5 above, we made a choice of which edge of weight 3 to add to the subgraph (that would eventually become a spanning tree). There are two other possible minimum spanning trees (each of which has total weight 17) that correspond to the other options for picking two of the three edges of weight 3. These are shown below.

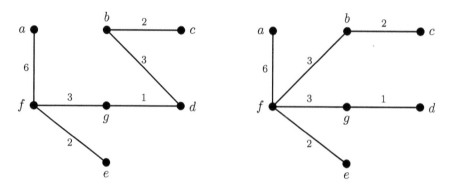

Perhaps the most surprising aspect of Kruskal's Algorithm is the process you would like to take (namely picking the small edges) also guarantees a minimum spanning tree. Looking back at the example above, when we skipped over an edge (say bg of weight 5), we did so because including it would create a cycle. This means a path between the endpoints of that edge (say b and g) must already exist and the other edges along that path must each be of weight no greater than the edge we skip over. In essence, Kruskal's Algorithm is focused on not creating cycles and eventually arrives at a connected subgraph. Conversely, if you think of finding a spanning tree as breaking cycles, then the largest edge on that cycle should never be chosen. This is the basis behind another algorithm, Reverse Delete, described in Exercise 4.8, that focuses on maintaining connectedness and eventually arrives at an acyclic subgraph.

Example 4.6 The Optos Cable Company is expanding its fiber optic network over the next few years. The company will need to lay new cable, but wishes to do so with minimal cost. A cost analysis estimates $15,000 per mile of cable installed. The distances (in miles) of required cable between any two towns is given in the table below. Determine an optimal network and its total cost.

	Mesa	Natick	Quechee	Rutland	Tempe	Vinton
Mesa	·	18	35	36	20	45
Natick	18	·	50	42	40	45
Quechee	35	50	·	41	25	19
Rutland	36	42	41	·	37	38
Tempe	20	40	25	37	·	15
Vinton	45	45	19	38	15	·

Solution: Apply Kruskal's Algorithm. The edges will be chosen directly from the chart above and a graph will be drawn. The chart indicates that the underlying graph is a complete graph, the drawing of which is omitted.

Step 1: Search the table for the smallest weight. This is Tempe-Vinton with weight 15.

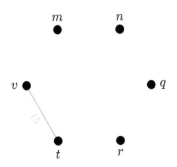

Step 2: The next smallest weight is 18 for Mesa-Natick.

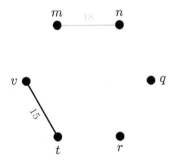

Step 3: Next is 19 for Quechee-Vinton.

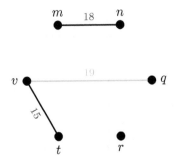

Step 4: The smallest weight is 20 for Mesa-Tempe. Since no circuit is created with its inclusion, we add the edge to the network.

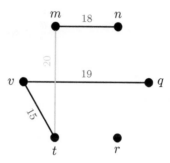

Step 5: The smallest weight remaining is 25 for Quechee-Tempe; however, this cannot be chosen since a circuit would be created between Quechee, Tempe, and Vinton.

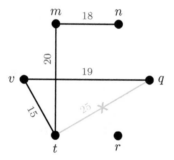

Skipping to the next smallest weight is Mesa-Quechee with 35. As above, we cannot choose this edge since it would create a circuit between Mesa, Quechee, Vinton, and Tempe.

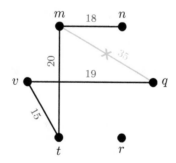

The next smallest edge is 36 for Mesa-Rutland; this is a valid edge. The network now spans the original graph and no further edges need to be added.

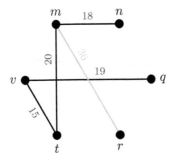

Output: Spanning tree of total length 108 miles and estimated cost $1,620,000.

Look back at the graphs obtained during each step of Kruskal's algorithm in Examples 4.5 and 4.6. Notice that the subgraphs obtained are acyclic but may not be connected; these subgraphs are forests. This is not problematic in a purely mathematical sense, but could pose complications if the original graph modeled a real world scenario. In Example 4.6, the Optos Cable Company needed to expand their fiber-optic network. Would it make sense for them to lay cable between two cities that are not connected to their hub, when this implies the fiber optic service could not yet be activated?

Prim's Algorithm

The algorithm described below is widely known as Prim's Algorithm, named for the American mathematician and computer scientist Robert C. Prim. Prim worked closely with Kruskal at Bell Laboratories and published this algorithm in 1957 (one year after the publication of Kruskal's Algorithm). However, it was originally discovered in 1930 by the Czech mathematician Vojtěch Jarnik and also republished in 1959 by Dijkstra (who should be familiar from Chapter 3).[19]

Prim's Algorithm contrasts from Kruskal's in that the structure obtained in each step is itself a tree. By the end of the process, a spanning tree will be found. It begins by denoting a starting vertex for the tree, similar to the root from Example 4.3.

Prim's Algorithm

Input: Weighted connected graph $G = (V, E)$.

Steps:

1. Let v be the root. If no root is specified, choose a vertex at random. Highlight it and add it to $T = (V', E')$.

2. Among all edges incident to v, choose the one of minimum weight. Highlight it. Add the edge and its other endpoint to T.

3. Let S be the set of all edges with exactly endpoint from $V(T)$. Choose the edge of minimum weight from S. Add it and its other endpoint to T.

4. Repeat step (3) until T contains all vertices of G, that is $V(T) = V(G)$.

Output: Minimum spanning tree T of G.

Similar to Dijkstra's Algorithm from Chapter 3, we will consider vertices adjacent to previously chosen vertices. Unlike Dijkstra's Algorithm, however, we are not concerned with weights along a path but rather the total weight of all edges chosen.

Example 4.7 Use Prim's algorithm to find a minimum spanning tree for the graph given in Example 4.5.

Solution: As before, previously chosen edges will be in black and the newly chosen edge in blue.

Step 1: Since no root was specified, we choose a as the starting vertex.

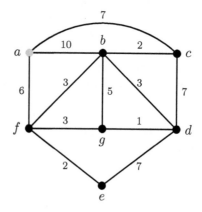

Step 2: We consider the edges incident to a, namely ab, ac and af. These are shown in blue in the graph on the left. The edge of least weight is af. This is added to the tree, shown in blue on the right.

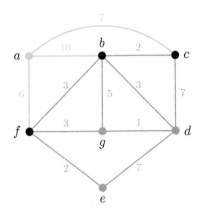
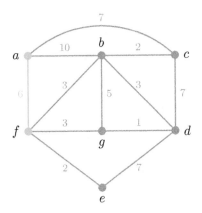

Step 3: The set S consists of edges with one endpoint as a or f. These are shown in blue on the graph to the left. The edge of minimum weight from these is ef. This is added to the tree, as shown on the right.

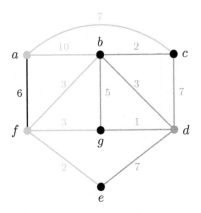
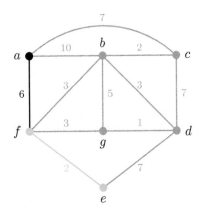

Step 4: The new set S consists of edges with one endpoint as a, e, or f, as shown in blue on the left. The next edge added to the tree could either be fg or fb. We choose fg.

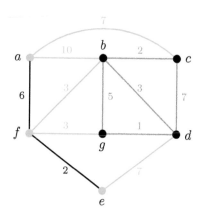
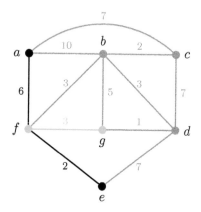

Step 5: We consider the edges where exactly one endpoint is from a, e, f, or g. The next edge to add to the tree is dg.

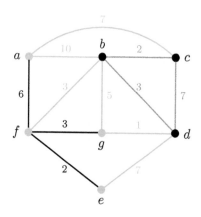
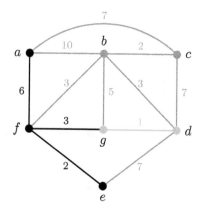

Step 6: The edges to consider must have exactly one endpoint from a, d, e, f, or g. Note that de is no longer available since both endpoints are already part of the tree (and its addition would create a cycle). There are two possible minimum weight edges, bf or bd. We choose bf.

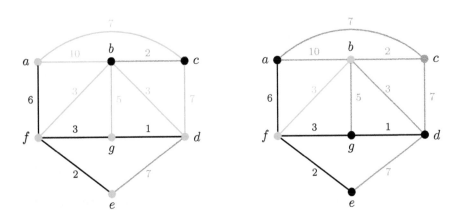

Step 7: The only edges we can consider are those with one endpoint of c since this is the only vertex not part of our tree. The edge of minimum weight is bc.

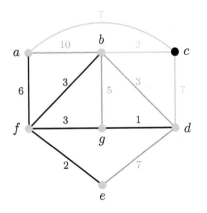
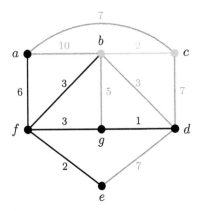

Output: A minimum spanning tree of total weight 17.

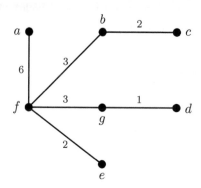

Unlike Kruskal's Algorithm, Prim's provides a plan for how to build the spanning tree. As noted earlier, this is beneficial for use in real world problems that have been modeled mathematically. One such example is outlined below.

Example 4.8 Use Prim's Algorithm to find a minimum spanning tree if Optos Cable Company from Example 4.6 must expand its fiber optic network from its headquarters in Quechee.

Solution: As with Example 4.6, the drawing of the underlying complete graph is omitted. The possible choices are shown in gray and the edge of minimum weighted is highlighted in blue.

Step 1: First look for the smallest edge from Quechee. This is Quechee-Vinton of weight 19.

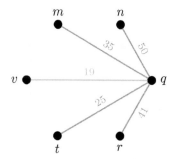

Step 2: Next search for the edge of smallest weight with exactly one endpoint as Quechee or Vinton. Pick Vinton-Tempe with weight 15.

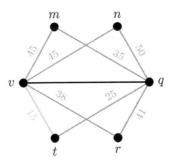

Step 3: We now need edges with exactly one endpoint from Quechee, Tempe, or Vinton. The smallest is Tempe-Mesa with weight 20.

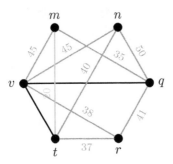

Step 4: The edges we consider must have exactly one endpoint from Mesa, Quechee, Tempe, or Vinton. The smallest weight is 18 for Mesa-Natick.

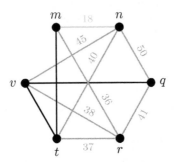

Step 5: There is only one vertex left to add to the spanning tree, and so we must find the smallest edge to Rutland, which is Mesa-Rutland with weight 36.

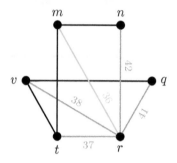

Output: A minimum spanning tree with total weight 108.

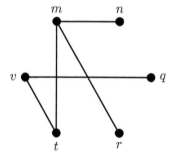

Both minimum spanning tree algorithms described in this chapter are efficient and optimal, and result in roughly the same computation requirements. It should be noted that many other algorithms exist and the study of minimum spanning trees did not originate with Kruskal and Prim. In fact, both mathematicians cited the work of Otakar Borůvka, a Czech mathematician who is credited with the first minimum spanning tree algorithm from 1926. Borůvka's Algorithm, which can be found in Exercise 4.14, is slightly more complex and requires all edge weights to be unique, though is better suited for parallel computing.

4.3 Shortest Networks

In Example 4.6 Optos Cable Company is building a fiber-optic network to connect their customers in six cities. The junction point of two stretches of cable always appeared within one of the cities. From a truly practical standpoint, this may not always be optimal. By adding new vertices in strategic locations, the total weight of the network may be reduced; however, the network created likely will not be a spanning tree of the original graph. So as not to confuse it with the Minimum Spanning Tree Problem from the previous section, we refer to this as the *Shortest Network Problem*.

The Shortest Network Problem is at the intersection of graph theory and geometry, and we will be using some geometric properties and results in our search for a solution. One item of caution before we delve much deeper into Shortest Networks: up to this point, we have not been concerned with how a graph is drawn, but rather making the visualization as easy to understand as possible. For the remainder of this section, it is imperative that edge lengths are drawn to scale since the angle created by the edges will play a large role in determining where to place shortcuts.

We begin with the smallest example that can have any interest — a network with three vertices. Consider the two triangles shown below, with lengths of the sides indicated. To find a shortest network, we begin by finding a minimum spanning tree, which consists of two of the three legs of the triangle. For T_1, the two edges of length 10.3 would be chosen for a total tree weight of 20.6; for T_2, any two of the three edges could be chosen for a tree of weight 20. But could we do better than a Minimum Spanning Tree?

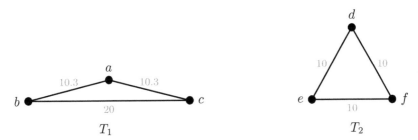

A similar question was posed by the 17th century French mathematician Pierre de Fermat in a letter to Evangelista Torricelli, an Italian physicist and mathematician. In his letter, Fermat challenged Torricelli to find a point that minimizes the distance to each of the vertices in a triangle; such a point is called the *Fermat point* of a triangle.

Definition 4.3 A *Fermat point* for a triangle is the point p so that the total distance from p to the vertices of the triangle is minimized. Each of the three angles formed by these segments measures 120°.

Notice triangle T_1 appears rather flat in comparison with triangle T_2, which is an equilateral triangle. Thinking of visiting the third vertex (a or d) as a detour along the straight line path between the other two (b to c or e to f), we find this detour is much closer to the straight line distance in the flat triangle (20.6 vs. 20) than in the equilateral triangle (20 vs. 10). This implies there is less room for improvement by adding a new vertex. In fact, Torricelli proved that triangles with a rather large angle (making the triangle appear flatter) prohibits the existence of a shorter network than the minimum spanning tree. His result is summarized in Theorem 4.4 below.

Theorem 4.4 Given three points and a triangle T formed by these points, the Shortest Network connecting the three points will either be

1. the two shortest sides of T provided T has one angle of at least 120°; or

2. the three segments connecting the Fermat point for T to the original three vertices of T.

In addition to determining the existence question (when does a triangle have a Fermat point), Torricelli also answered the construction question (how to find the Fermat point) and in fact gave more than one solution. The algorithm described below is a modified form of one of these solutions.

Torricelli's Construction

Input: Triangle $T = \triangle abc$ where all angles measure less than 120°.

Steps:

1. Along edge ab of T construct an equilateral triangle using that edge and a new point x that is on the opposite side of the edge as c.

2. Repeat Step (1) for the other two edges of T, introducing new points y and z across from b and a, respectively.

3. Join x and c, y and b, and z and a by a line segment.

4. The point of concurrency (intersection point of three lines) is the Fermat point p for T.

5. The shortest network is the line segments joining each of the original vertices, $a, b,$ and c, with p.

Output: Fermat point p and shortest network connecting $a, b,$ and c.

Note that this construction does not require anything other than a ruler and a compass (or protractor). The example below demonstrates how to apply Torricelli's Construction to a triangle in which all the angles measure less than 120°.

Example 4.9 Use Torricelli's Construction to find the Fermat point and Shortest Network for the triangle below.

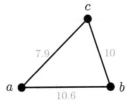

Solution:

Step 1: Form an equilateral triangle off edge ab with new point x on the opposite side of ab as c.

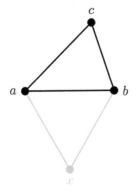

Step 2: Repeat for edge ac.

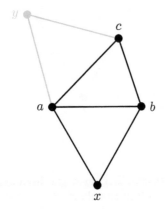

Step 3: Repeat for edge bc

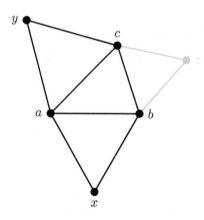

Step 4: Join the new vertices to their opposite vertex.

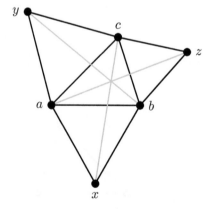

Step 5: Find the point of concurrency p and highlight the edges from p to each of the original vertices to find the shortest network.

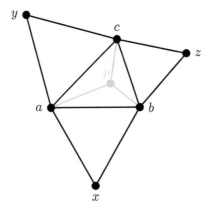

Output: The shortest network consists of the edges from p back to each of the original vertices.

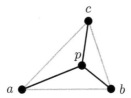

The beauty of Torricelli's Construction is the ease at which the Fermat point and Shortest Network are found. There is slight difficulty in determining the length of the line segments from the Fermat point to the original vertices. Since we will always be concerned with the network itself, we have the following formula for finding the length of the network connecting 3 points.

If l denotes the length of the shortest network connecting vertices of a triangle having angles that measure less than $120°$ each and sides of lengths $a, b,$ and c, then

$$l = \sqrt{\frac{a^2 + b^2 + c^2}{2} + \frac{\sqrt{3}}{2}\sqrt{2a^2b^2 + 2a^2c^2 + 2b^2c^2 - (a^4 + b^4 + c^4)}}$$

Note that l gives the *total* length of the three legs introduced with the Steiner point.

Example 4.10 Find the length of the network obtained in Example 4.9 and compare it to the Minimum Spanning Tree for the graph.

Solution: Using the formula above with $a = 7.9, b = 10$ and $c = 10.6$, we estimate the shortest network to have length

$$l = \sqrt{\frac{274.77}{2} + \frac{\sqrt{3}}{2}\sqrt{48978.7752 - 26519.7777}} \approx 16.345$$

The Minimum Spanning Tree has a total length of 17.9 (choose the two smallest edges) and so the shortest network saves 1.555, or 8.69%.

In the instances when the exact length of the segments are needed, the open source (and free!) software package *GeoGebra* makes the process of finding Fermat points and the lengths of individual segments much easier. It is recommended for use whenever more accurate calculations are required. See the Appendix for more information on using GeoGebra for finding Fermat points.

Steiner Trees

As the example above demonstrates, finding a shortest network for a situation with only three locations can be done with very little advanced mathematics. When the original graph contains more than three points, we encounter a bit more difficulty in finding the shortest network. We will continue to investigate the optimization question (what is the shortest network between a given set of points) though our improvement on the minimum spanning tree may not always be optimal. In situations where we have more than three original points, the short-cut points are no longer called Fermat points, but rather Steiner points.

Definition 4.5 For a graph G, a **Steiner point** is a new point p added to the graph that has vertex degree of 3 where the three edges incident to p form $120°$ angles. A **Steiner tree** is a tree that only consists of Steiner points and the original vertices of G.

Note that when describing a Steiner point, we make use of two different notions of degree: vertex degree, as we have primarily been using throughout this text, and angle measure in degrees, as is more familiar from geometry courses.

A Steiner point is similar to a Fermat point in that it is added to graph to find the shortest network connecting the original vertices. It has been shown that finding a shortest network amounts to finding a minimum Steiner tree. This problem is named for the 19th century Swiss mathematician Jakob Steiner who mainly studied geometry. Though the Steiner Tree Problem sounds fairly simple, it is in fact among a class of problems known as *NP*-Hard which can be thought of as problems for which solutions can be verified quickly, though finding a solution can be quite hard. For further information on *NP*-Hard and other computation classification of problems, see Section 7.1.

Consider the simplest example with more than three vertices (since the three-vertex problem has already been answered) — a graph with four vertices that lie on a square as shown below on the left. A minimum spanning tree consists of three of the four edges of length 10. One such example is shown below on the right.

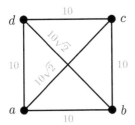 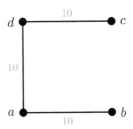

We can think of phantom triangles existing in this network, by adding in one of the removed diagonals and thus allowing us to find a Fermat point for this triangle. The example below uses the triangle formed by a, c and d.

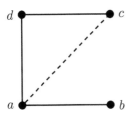 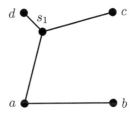

If we do this again, working from the triangle using the new point s_1 and the previously unconsidered vertex (b), we can find another Fermat point for this triangle. This produces a network of length roughly 27.852, an improvement of 2.148 over the minimum spanning tree.

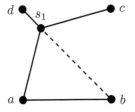 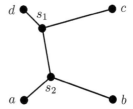

However, these two points are not Steiner points, since they do not both have edges that form 120° angles and so this tree, although shorter than the minimum spanning tree, is not the minimum Steiner tree. The minimum Steiner tree is given below and has a total length of 27.321.

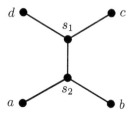

We will not be concerned with finding the minimum Steiner tree, but rather use the ideas behind a Steiner tree to find a shorter network than a minimum spanning tree, if this is possible. The use of *GeoGebra* is recommended for the remainder of this section.

Steiner Network Method

1. Find the minimum spanning tree of the network.

2. Form a triangle from two existing edges of the minimum spanning tree. If all angles of this triangle measure less than 120°, find the Fermat point.

3. Update the network by removing the two edges from the minimum spanning tree used in Step (2) and adding new edges to the Fermat point.

4. Repeat Steps (2) and (3) until all possible triangles have been considered.

The example below describes how to use the Steiner Network Method in finding a shorter network than the minimum spanning tree. All calculations were done through the use of GeoGebra.

Example 4.11 Optos Cable Company has hired you to find the shortest network connecting seven locations for their newest fiber optic network expansion. You have already found the minimum spanning tree for these locations, as shown below, where the edge labels represent the distance in miles. Find a shorter network using the Steiner Network Method and determine the savings if the cost of cable installation is $15,000 per mile.

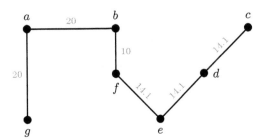

Solution: We begin with the vertices a, b, and g since the angle at a is $90°$. Using Torricelli's Construction, we get an improved network shown below.

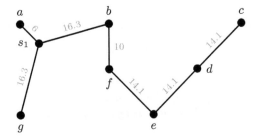

We next work with vertices f, e, and d since these also have a $90°$ at e. Using Torricelli's Construction, we get an improved network shown below.

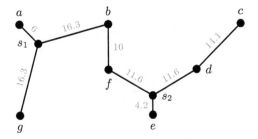

At this point, due to angle measures, there is only one location that can be improved. This occurs between vertices b, f and s_1 since the triangle formed has angles of $75°, 70°$ and $35°$. The updated network is shown below.

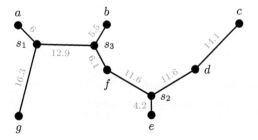

The shorter network shown above has a total length of 88.3 versus 92.3 miles for the minimum spanning tree. The cost savings is $60,000 by using the network above.

Although we did not fully answer the optimization question for networks containing more than three points, using the Steiner Network Method provides a quick and simple procedure for finding locations for improvements to the minimum spanning tree.

4.4   Traveling Salesman Problem Revisited

In Chapter 2, we discussed multiple algorithms for finding an approximate solution to a Traveling Salesman Problem. Within each of these, we allowed the weight on an edge to represent either cost, distance, or time. The metric Traveling Salesman Problem (mTSP) only considers scenarios where the weights satisfy the ***triangle inequality***; that is, for a weighted graph $G = (V, E, w)$, given any three vertices x, y, z we have

$$w(xy) + w(yz) \geq w(xz)$$

The triangle inequality is named to reference a well-known fact in geometry that no one side of a triangle is longer than the sum of the other two sides. When the weight function is modeling distance, we are within the mTSP realm; when the weight function models cost or time, we may or may not be in a scenario that satisfies the triangle inequality.

Minimum spanning trees, and the algorithms used to find such subgraphs, can be used to find an approximate solution to a metric Traveling Salesman Problem. The algorithm below combines three ideas we have studied so far: Eulerian circuits, Hamiltonian cycles, and minimum spanning trees. A minimum spanning tree is modified by duplicating every edge, ensuring all vertices have even degree and allowing an Eulerian circuit to be obtained. This circuit is then modified to create a Hamiltonian cycle. Note that this procedure is guaranteed to work only when the underlying graph is complete. It may still find a proper Hamiltonian cycle when the graph is not complete, but cannot be guaranteed to do so.

mTSP Algorithm

Input: Weighted complete graph K_n, where the weight function w satisfies the triangle inequality.

Steps:

1. Find a minimum spanning tree T for K_n.

2. Duplicate all the edges of T to obtain T^*.

3. Find an Eulerian circuit for T^*.

4. Convert the Eulerian circuit into a Hamiltonian cycle by skipping any previously visited vertex (except for the starting and ending vertex).

5. Calculate the total weight.

Output: Hamiltonian cycle for K_n.

The example below is similar to those from Chapter 2, except the distances shown satisfy the triangle inequality. Recall that to find an optimal Hamiltonian cycle on a graph with 6 vertices, we would need to calculate all 60 possible Hamiltonian cycles.

Example 4.12 Nour must visit clients in six cities next month and needs to minimize her driving mileage. The table below lists the driving distances between these cities. Use the mTSP Algorithm to find a good plan for her travels if she must start and end her trip in Philadelphia. Include the total distance.

	Boston	Charlotte	Memphis	New York	Philadelphia	D.C.
Boston	.	840	1316	216	310	440
Charlotte	840	.	619	628	540	400
Memphis	1316	619	.	1096	1016	876
New York City	216	628	1096	.	97	228
Philadelphia	310	540	1016	97	.	140
Washington, D.C.	440	400	876	228	140	.

Solution: The details for finding a minimum spanning tree and an Eulerian circuit will be omitted (You are encouraged to work through these!). In addition, city names will be represented by their first letter.

Step 1: A minimum spanning tree for the six cities is given below.

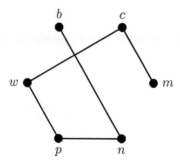

Step 2: Duplicate all the edges of the tree above.

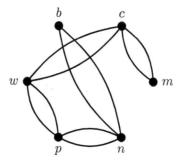

Step 3: Find an Eulerian circuit starting at p. The circuit shown below is $p\,n\,b\,n\,p\,w\,c\,m\,c\,w\,p$.

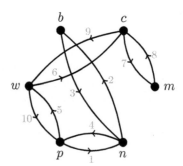

Step 4: We follow the Eulerian circuit from Step 3 until we reach vertex b. Since we are looking for a Hamiltonian cycle, we cannot repeat vertices and so we cannot return to n. The next vertex along the circuit that has not been previously visited is w.

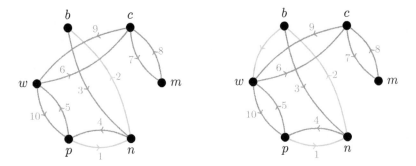

We follow the circuit again until m is reached. Again, we cannot return to c and at this point we must return to p since all other vertices have been visited.

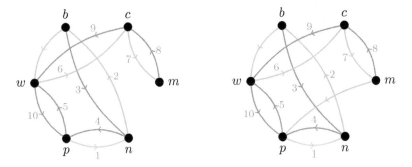

Output: The final Hamiltonian cycle is shown below (Philadelphia – New York – Boston – Washington, D.C. – Charlotte – Memphis – Philadelphia) with a total weight of 2788 miles.

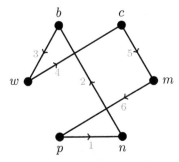

In the example above, our minimum spanning tree had a total weight of 1472, implying that the worst possible Hamiltonian cycle that can arise from it will have weight at most two times that, 2944, due to the doubling of the edges. In fact, for this scenario, the optimal cycle has total weight of 2781 miles, making the result of 2788 from the mTSP within 7 miles of optimal, or off by a relative error of only 0.25%!

In general, the mTSP Algorithm performs on par with the approximation algorithms from Chapter 2. A modification of this algorithm, Christofides' Algorithm, appears in the exercises in the next chapter as it makes use not only of minimum spanning trees but also the topic of Chapter 5, graph matchings.

4.5   Exercises

4.1 For each of the graphs described below, determine if G is (i) definitely a tree, (ii) definitely not a tree, or (iii) may or may not be a tree. Explain your answer or demonstrate with a proper graph.

  (**a**) G has 10 vertices and 11 edges.
  (**b**) G has 10 vertices and 9 edges.
  (**c**) G is connected and every vertex has degree 1 or 2.
  (**d**) There is exactly one path between any two vertices of G.
  (**e**) G is connected with 15 vertices and 14 edges.
  (**f**) G is connected with 15 vertices and 20 edges.
  (**g**) G has two components, each with 9 vertices and 8 edges.

4.2 Find a spanning tree for each of the graphs below.

(**a**)   (**b**)

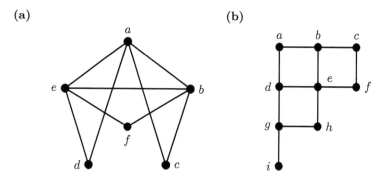

(**c**)

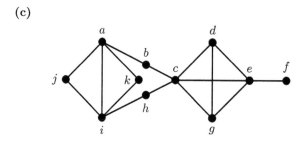

(d)

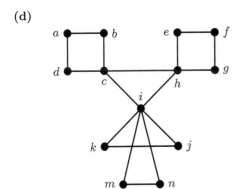

4.3 Find a minimum spanning tree for each of the graphs below using (i) Kruskal's Algorithm and (ii) Prim's Algorithm.

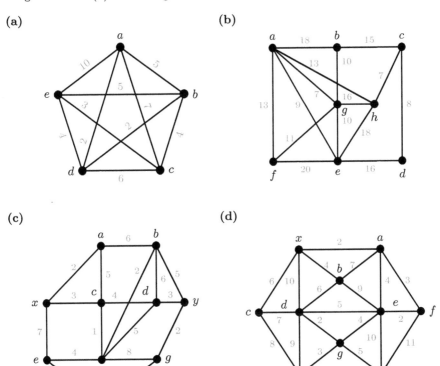

(e)

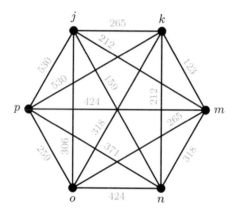

4.4 Find a minimum spanning tree for the graph represented by the table below.

	a	b	c	d	e	f	g
a	·	5	7	8	10	3	11
b	5	·	2	4	1	12	7
c	7	2	·	6	7	5	4
d	8	4	6	·	2	10	12
e	10	1	7	2	·	6	9
f	3	12	5	10	6	·	15
g	11	7	4	12	9	15	·

4.5 Kruskal's Algorithm and Prim's Algorithm are both written with a connected graph as an input. Determine how each of these would perform if the input was a disconnected graph.

4.6 How would you modify Kruskal's and Prim's algorithm if a specific edge must be included in the spanning tree? Would the resulting tree be a minimum spanning tree? Explain your answer.

4.7 Optos Cable Company from Example 4.8 has completed the first two steps of their fiber optic expansion (so edges qv and vt have been created). Due to other infrastructure construction, the mileage need for cable between some of the cities has changed. Use the edges already constructed and the new weights shown below to finish their network expansion. Find the total weight and cost of building their network.

	Mesa	Natick	Quechee	Rutland	Tempe	Vinton
Mesa	·	29	32	25	34	45
Natick	29	·	50	46	45	51
Quechee	32	50	·	41	28	19
Rutland	25	46	41	·	40	35
Tempe	34	45	28	40	·	15
Vinton	45	51	19	35	15	·

4.8 The Reverse Delete Algorithm finds a minimum spanning tree by deleting the largest weighted edges as long as you do not disconnect the graph. In essence, it is Kruskal's Algorithm in reverse. Verify that Reverse Delete produces the same minimum spanning trees for the graphs from Examples 4.5 and 4.6.

4.9 Under what circumstances would Reverse Delete be a better choice than Kruskal's Algorithm? Under what circumstance would Kruskal's be a better choice? Explain your answer.

For Problems 4.10 and 4.11, a protractor and ruler or the computer program GeoGebra (using the vertex coordinates given below the graph) is required.

4.10 For each of the following triangles, perform Torricelli's Construction to find the Fermat point. It is recommended that you trace the triangles onto another piece of paper and use a protractor and ruler.

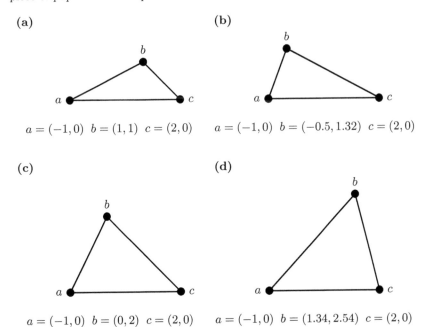

(a)

$a = (-1, 0)$ $b = (1, 1)$ $c = (2, 0)$

(b)

$a = (-1, 0)$ $b = (-0.5, 1.32)$ $c = (2, 0)$

(c)

$a = (-1, 0)$ $b = (0, 2)$ $c = (2, 0)$

(d)

$a = (-1, 0)$ $b = (1.34, 2.54)$ $c = (2, 0)$

4.11 Each of the following graphs represents the spanning tree for a network. Use the Steiner Network Method to find a shorter network.

(a)

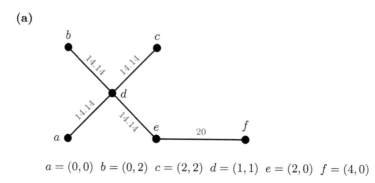

$$a = (0,0) \quad b = (0,2) \quad c = (2,2) \quad d = (1,1) \quad e = (2,0) \quad f = (4,0)$$

(b)

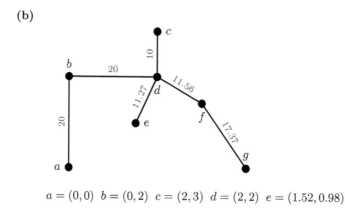

$$a = (0,0) \quad b = (0,2) \quad c = (2,3) \quad d = (2,2) \quad e = (1.52, 0.98)$$
$$f = (3, 1.42) \quad g = (4,0)$$

(c)

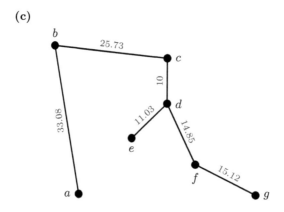

$$a = (0,0) \quad b = (-0.56, 3.26) \quad c = (2,3) \quad d = (2,2) \quad e = (1.2, 1.24)$$
$$f = (2.64, 0.66) \quad g = (4,0)$$

4.12 Nour must visit clients in six cities next month and needs to minimize her driving mileage. The table below lists the distances between these cities. Use the mTSP Algorithm to find a good plan for her travels if she must start and end her trip in Dallas. Include the total distance.

	Austin	Dallas	El Paso	Fort Worth	Houston	San Antonio
Austin	·	182	526	174	146	74
Dallas	182	·	568	31	225	253
El Paso	526	568	·	537	672	500
Fort Worth	174	31	537	·	237	241
Houston	146	225	672	237	·	189
San Antonio	74	253	500	241	189	·

4.13 The weight of the edges of the graph in Exercise 4.3(e) satisfies the triangle inequality. Apply the mTSP Algorithm to find a Hamiltonian cycle and compare it to those found from Chapter 2 (Exercise 2.8(e)).

Projects

4.14 In Section 4.2, we studied two different algorithms for finding a minimum spanning tree. As mentioned earlier, both Kruskal and Prim cited the work of the Czech mathematician Otakar Borůvka. Below is a description of his algorithm, first published in 1926.

Borůvka's Algorithm

Input: Weighted connected graph $G = (V, E)$ where all the weights are distinct.

Steps:

1. Let T be the forest where each component consists of a single vertex.

2. For each vertex v of G, add the edge of least weight incident to v to T.

3. If T is connected, then it is a minimum spanning tree for G. Otherwise, for each component C of T, find the edge of least weight from a vertex in C to a vertex not in C. Add the edge to T.

4. Repeat Step (3) until T has only one component, making T a tree.

Output: A minimum spanning tree for G.

Apply Borůvka's Algorithm to the following two graphs. Use either Kruskal's Algorithm or Prim's Algorithm to verify that Borůvka's Algorithm found a minimum spanning tree.

(a)

(b)

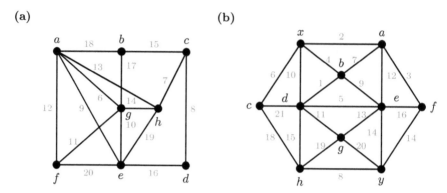

4.15 Come up with a business that needs to find a shortest network. Name the business and describe why they are working on this problem. Make sure to include a good reason why finding a shortest network is necessary for their business. Using the table at the end of Chapter 2 (on pages 79 and 80), choose 7 locations and draw the weighted graph. Apply one of the algorithms from this chapter to find the minimum spanning tree. Draw your minimum spanning tree on a map showing all 7 locations, and use the Steiner Network Method to find an improvement from the minimum spanning tree, or argue why one does not exist. Find locations for improvements.

Chapter 5

Matching

This chapter and the next one are big departures from the previous four. Up to this point we have mainly used graphs to model routing problems (circuits, cycles, paths) or exhaustive problems (spanning trees, shortest networks, Eulerian and Hamiltonian tours). While are still concerned with maximizing or minimizing items, the graph models we will use focus on resource requirements for a specific problem. This chapter investigates the optimization of pairings through the use of edge-matchings within a graph, more commonly known as a matching.

Definition 5.1 Given a graph $G = (V, E)$, a *matching* M is a subset of the edges of G so that no two edges share an endpoint. When two edges do not share an endpoint, we call them *independent* edges. The size of a matching, denoted $|M|$, is the number of edges in the matching.

The most common application of matchings is the pairing of people, usually described in terms of marriages. Other applications of a graph matching are task assignment, distinct representatives, and roommate selection. Consider the following scenario:

> A company receives a last minute order that needs to be filled in time to be shipped. None of the tasks rely on each other but no person has enough time to complete more than one task. In addition, most employees are only qualified to complete some of the tasks. Determine the best way to assign tasks to employees so the order can be completed in time.

Although this scenario resembles those in the Project Scheduling section of Chapter 3, the slight change in details makes the digraph model approach a poor fit. Instead, bipartite graphs will be utilized to model a matching problem. The last section of the chapter will discuss matchings in graphs that are not bipartite.

5.1 Bipartite Graphs

Bipartite graphs first appeared in Example 1.4 when considering room restrictions for various student organizations. As noted earlier, bipartite graphs are often used to model interactions between two distinct types of groups.

Definition 5.2 A graph $G = (V, E)$ is **bipartite** if the vertices can be partitioned (or split) into two sets, X and Y, so that X and Y have no vertices in common, every vertex appears in either X or Y, and every edge has exactly one endpoint in X and the other endpoint in Y. We denote this by $G = (X \cup Y, E)$.

The graph shown below is bipartite. The vertices have been partitioned into sets $X = \{a, b, c\}$ and $Y = \{d, e, f\}$.

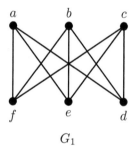

G_1

At the beginning of Chapter 1, we discussed how a graph can be drawn in more than one way. The drawing of G_1 above emphasizes the bipartite nature of the graph, with the two sets of the partition drawn on a different level. However, a graph can be bipartite without being drawn in this form.

Determining if a graph is bipartite is much easier than it may first appear. If you are traveling along a path or cycle in a graph, the vertices will need to alternate between the two parts of the vertex set, such as a vertex from X, then Y, then X, etc. So if a cycle exists in the graph, it must have even length since otherwise two vertices along the cycle would come from the same part. This result is surprisingly useful and restated in the theorem below.

Theorem 5.3 A graph G is bipartite if and only if there are no odd cycles in G.

In practice, we usually search for odd cycles within a graph and if we cannot find any, we try to redraw the graph to emphasize that it is bipartite. Note that a bipartite graph can have multi-edges (and so need not be simple) but cannot have loops (since these are odd cycles of length 1).

Example 5.1 Which of the following graphs are bipartite?

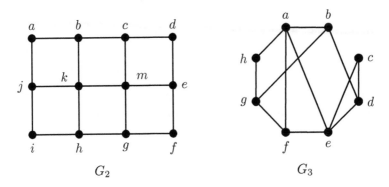

Solution:

- G_2 is bipartite. We can partition the vertices as $X = \{a, c, e, g, i, k\}$ and $Y = \{b, d, f, h, j, m\}$ and the drawing below emphasizes this vertex partition.

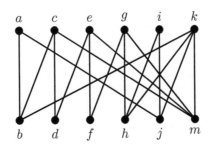

- G_3 is not bipartite since there are odd cycles, such as the 5-cycle $aefgha$ (highlighted below) or the 3-cycle $cdec$. Can you find any other odd cycles?

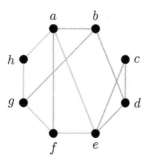

The graph G_1 from above is a special type of bipartite graph since every vertex from X has exactly one edge to every vertex from Y. These are called *complete bipartite graphs* to parallel the notion of a complete graph.

Definition 5.4 A simple bipartite graph $G = (X \cup Y, E)$ is a ***complete bipartite graph*** if every vertex in X is adjacent to every vertex in Y. If the size of X is m and the size of Y is n, then we write $K_{m,n}$.

With this notation, G_1 above is named $K_{3,3}$. Further examples of complete bipartite graphs appear in Section 5.3 and in the exercises.

5.2   Matching Terminology and Strategies

Matching problems often (though not always) make use of bipartite graphs since the items being matched are usually of two distinct types. The example below illustrates how to model a task assignment problem as a matching in a bipartite graph.

Example 5.2 The Vermont Maple Factory just received a rush order for 6 dozens of maple cookies, 3-dozen bags of maple candy, and 10-dozen bottles of maple syrup. Some employees have volunteered to stay late tonight to help finish the orders. In the chart below, each employee is shown along with the jobs for which he or she is qualified. Draw a graph to model this situation and find a matching.

Employee	Task	
Dan	Making Cookies	Bottling Syrup
Jeff	Labeling Packages	Bottling Syrup
Kate	Making Candy	Making Cookies
Lilah	Labeling Packages	
Tori	Labeling Packages	Bottling Syrup

Solution: Model using a bipartite graph where X consists of the employees and Y consists of the tasks. We draw an edge between two vertices a and b if employee a is capable of completing the task b.

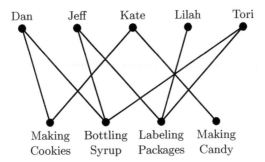

A matched edge, which is shown in blue below, represents the assignment of a task to an employee. One possible matching is shown below.

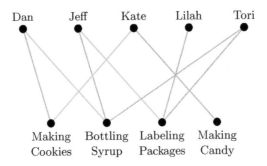

With any matching problem, you should ask yourself what is the important criteria for a solution and how does that translate to a matching. In Example 5.2, is it more important for each employee to have a task or for every task to be completed? We need a way to describe which vertices are the endpoint of a matched edge.

Definition 5.5 A vertex is *saturated* by a matching M if it is incident to an edge of the matching; otherwise, x is called *unsaturated*.

The matching displayed in Example 5.2 has saturated vertices (Dan, Jeff, Kate, Making Cookies, Bottling Syrup, and Labeling Packages) representing the three tasks that will be completed by the three employees. The unsaturated vertices (Making candy, Lilah, and Tori) represent the tasks that are not assigned or the employees without a task assignment. Is this a good matching? No; some parts needed to complete the order are not assigned and so the order will not be fulfilled. When searching for a matching in a graph, we need to determine what type of matching properly describes the solution.

Definition 5.6 Given a matching M on a graph G, we say M is

- *maximal* if M cannot be enlarged by adding an edge.

- *maximum* if M is of the largest size amongst all possible matchings.

- *perfect* if M saturates every vertex of G.

- an *X-matching* if it saturates every vertex from the collection of vertices X (a similar definition holds for a Y matching).

Note that a perfect matching is automatically maximum and a maximum matching is automatically maximal, though the reverse need not be true. Consider the two graphs below, with a matching shown in blue. The matching on the left is maximal as no other edges in the graph can be added since the remaining edges require the use of a saturated vertex (either a for edges ac and ae or d for edge bd). The matching on the right is maximum since there is no way for a matching to contain three edges (since then vertex a would have two matched edges incident to it). In addition, the matching on the right is an X matching if we define $X = \{a, b\}$. Finally, neither matching is perfect since not every vertex is saturated.

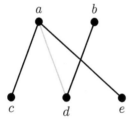   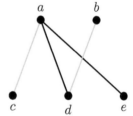

Maximal Matching   Maximum Matching

Depending on the scenario, the existence question we are examining translates to a search for a perfect matching or an X-matching. However, when neither of these can be found, we need a good explanation as to the size of a maximum matching, which will appear later in Section 5.2. For now, we will find a solution to the rush order at the Vermont Maple Factory.

Example 5.3 Determine and find the proper type of matching for the Vermont Maple Factory from Example 5.2.

Solution: Since we need the tasks to be completed but do not need every employee to be assigned a task, we must find an X-matching where X consists of the vertices representing the tasks. An example of such a matching is shown below.

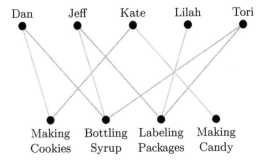

Note that all tasks are assigned to an employee, but not all employees have a task (Lilah is not matched with a task). In addition, this is not the only matching possible. For example, we could have Lilah labeling packages and Tori bottling syrup with Jeff having no task to complete.

Example 5.3 above illustrates that many matchings can have the same size. Thus when finding a maximum matching, it is less important which people get paired with a task than it does that we make as many pairings possible. Section 5.3 addresses the scenario when we not only need to find a maximum matching, but also one that fulfills additional requirements, such as preferences.

As with finding Eulerian circuits from Chapter 1, it is often quite clear how to form a matching in a small graph. However, as the size of the graph grows or the complexity increases, finding a maximum matching can become difficult. Moreover, once you believe a maximum matching has been found, how can you convince someone that a better matching does not exist? Consider the graph below with a matching shown in blue.

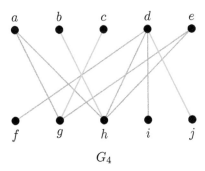

G_4

If you tried to adjust which edges appear in the matching to find one of larger size (try it!), you would find it impossible. But trial and error is a poor strategy for providing a good argument that you indeed have found a maximum matching. The theorem below addresses this through the use of a neighbor set. Recall that given a set of vertices S, the neighbor set $N(S)$ consists of all the vertices incident to at least one vertex from S.

Theorem 5.7 (Hall's Marriage Theorem) Given a bipartite graph $G = (X \cup Y, E)$, there exists an X-matching if and only if any collection S of vertices from X satisfies $|S| \leq |N(S)|$.

Looking back at the graph G_4 above, if we consider $S = \{f, i, j\}$, then $N(S) = \{d\}$ and so by Hall's Marriage Theorem there is no X-matching. In addition, since at most one of the vertices from S can be paired with d, we know the maximum matching can contain at most 3 edges. Since we found a matching with 3 edges, we know our matching is in fact maximum.

The theorem above is often referred to as Hall's Marriage Theorem since the early examples of matching were often described in terms of marriages between boys and girls within a small town. Note that the marriages considered in this text will be heterosexual marriages simply due to the need for two distinct groups that can only be matched with someone of a different type. Same sex marriages can be viewed as matchings in non-bipartite graphs, which appear in Section 5.4.

Example 5.4 In a small town there are 6 boys and 6 girls whose parents wish to pair into marriages where the only requirement is that a girl must like her future spouse (pretty low standards in my opinion). The table below lists the girls and the boys she likes. Find a pairing with as many marriages occurring as possible.

Girls	Boys She Likes			
Opal	Henry	Jack		
Penny	Gavin	Isa	Henry	Lucas
Quinn	Henry	Jack		
Rose	Kristof	Isa	Jack	
Suzanne	Henry	Jack		
Theresa	Gavin	Lucas	Kristof	

Solution: The information from the table will be displayed using a bipartite graph, where X consists of the girls and Y consists of the boys.

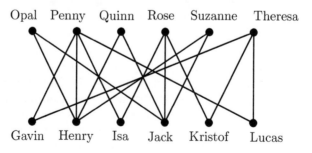

Notice that Opal, Quinn, and Suzanne all only like the same two boys (Henry and Jack), so at most two of these girls can be matched, as shown below.

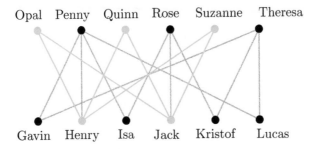

This means at most 5 marriages are possible; one such solution is shown below.

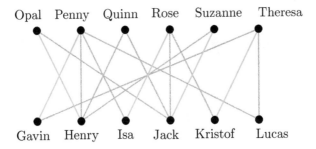

Hall's Marriage Theorem allows us to answer the existence question (when a graph has a perfect matching or X-matching) for bipartite graphs. When the answer to this question is negative, we move onto the optimization question (what is the size of a maximum matching). Hall's Marriage Theorem does not give a definitive answer about the size of a maximum matching but rather gives us the tools to reason why an X-matching does not exist. The next section uses a specific type of path and a collection of vertices as a way to determine the answer to the optimization question. In addition, an algorithm is described that finds a maximum matching within a bipartite graph, thus answering the construction question (how do we find a maximum matching?).

Augmenting Paths and Vertex Covers

Consider the graphs on page 156 showing the difference between a maximal and maximum matching, which are reproduced as graphs G_5 and G_6 below. Other than using trial and error to find a better matching, we need a way to determine if a matching is in fact maximum. We do this through the use of alternating and augmenting paths.

Definition 5.8 Given a matching M of a graph G, a path is called

- **M-*alternating*** if the edges in the path alternate between edges that are part of M and edges that are not part of M.

- **M-*augmenting*** if it is an M-alternating path and both endpoints of the path are unsaturated by M, implying both the starting and ending edges of the path are not part of M.

Both graphs below have alternating paths; for example, the path $c\,a\,d\,b$ is alternating in both graphs. However, this path is only augmenting in G_5 since both c and b are unsaturated by the matching. If we switch the edges along this path we get a larger matching. This switching procedure removes the matched edges and adds the previously unmatched edges along an augmenting path. Since the path is augmenting, the matching increases in size by one edge. Note that switching along the path $c\,a\,d\,b$ in G_5 produces the matching shown in G_6.

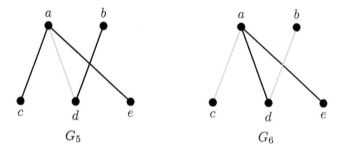

We previously discussed that the matching shown in G_6 is the maximum matching for that graph. Based on the discussion above, this should imply that no augmenting path exists (try it!), since otherwise we could switch along such a path to produce a larger matching. This result is stated in the theorem below and was first published by the French mathematician Claude Berge in 1959. Note that unlike Hall's Theorem, Berge's Theorem holds for both bipartite graphs and non-bipartite graphs.

Theorem 5.9 (Berge's Theorem) A matching M of a graph G is maximum if and only if G does not contain any M-augmenting paths.

Now that we understand how to determine if a matching is maximum (search for augmenting paths), we need a procedure or algorithm for the construction question (how do we find a maximum matching). The algorithm described below is closely related to the Hungarian Algorithm proposed by Harold Kuhn in 1955 [30]. He named this algorithm in honor of the work of the Hungarian mathematicians Dénes König and Jenő Egerváry whose largest contributions to graph matchings appear later in Theorem 5.11.

Augmenting Path Algorithm

Input: Bipartite graph $G = (X \cup Y, E)$.

Steps:

1. Find an arbitrary matching M.

2. Let U denote the set of unsaturated vertices in X.

3. If U is empty, then M is a maximum matching; otherwise, select a vertex x from U.

4. Consider y in $N(x)$.

5. If y is also unsaturated by M, then add the edge xy to M to obtain a larger matching M'. Return to Step 2 and recompute U. Otherwise, go to Step 6.

6. If y is saturated by M, then find a maximal M-alternating path from x using xy as the first edge.

   (a) If this path is M-augmenting, then switch edges along that path to obtain a larger matching M'; that is, remove from M the matched edges along the path and add the unmatched edges to create M'. Return to Step (2) and recompute U.

   (b) If the path is not M-augmenting, return to Step (4), choosing a new vertex from $N(x)$.

7. Stop repeating Steps (2) – (4) when all vertices from U have been considered.

Output: Maximum matching for G.

The arbitrary matching in Step 1 could be the empty matching (no edges are initially included in the matching), though in practice starting with a quick simple matching allows for fewer iterations of the algorithm. You should not spend much time trying to determine if the initial matching is maximum, but rather choose obvious edges to include. Also, even though Berge's Theorem holds for graphs that are not bipartite, this algorithm requires the input of a bipartite graph. A modification for general graphs is discussed in Section 5.4.

Example 5.5 Apply the Augmenting Path Algorithm to the bipartite graph below, where $X = \{a, b, c, d, e, f, g\}$ and $Y = \{h, i, j, k, m, n\}$, with an initial matching shown in blue.

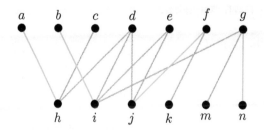

Solution:

Step 1: Define $U = \{c, d, e\}$ since these are the unsaturated vertices from X.

Step 2: Choose c. The only neighbor of c is h, which is saturated by M. Form an M-alternating path starting with the edge ch. This produces the path $c\,h\,a$ shown below, which is not augmenting.

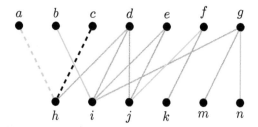

Step 3: Choose a new vertex from U, say d. Then $N(d) = \{h, i, j\}$. Below are the alternating paths originating from d.

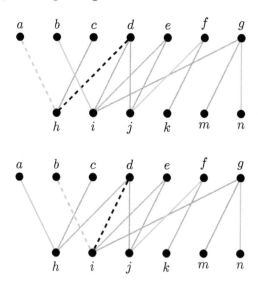

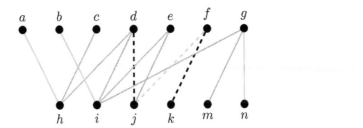

Note that the last path $(d\,j\,f\,k)$ is M-augmenting. Form a new matching M' by removing edge fj from M and adding edges dj and fk, shown below.

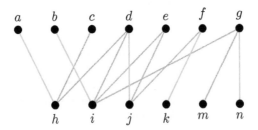

Step 4: Recalculate $U = \{c, e\}$. We must still check c since it is possible for the change in matching to modify possible alternating paths from a previously reviewed vertex; however, the path obtained is $c\,h\,a$, the same as from Step 2.

Step 5: Check the paths from e. The possible alternating paths are shown below.

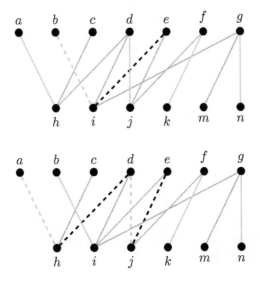

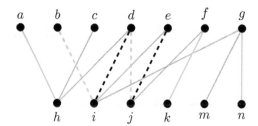

Note that none of these paths are augmenting. Thus no M'-augmenting paths exist in G and so M' must be maximum by Berge's Theorem.

Output: The maximum matching is $M' = \{ah, bi, dj, fk, gn\}$ as shown in Step 3.

The Augmenting Path Algorithm provides a method for not only finding a maximum matching, but also a reasoning why a larger matching does not exist since no augmenting paths exist at the completion of the algorithm. However, there are other ways to determine if a matching is maximum without the need to work through this algorithm. The simplest, and most elegant, is through the use of a specific set of vertices known as a vertex cover.

Definition 5.10 A *vertex cover* Q for a graph G is a subset of vertices so that every edge of G has at least one endpoint in Q.

Every graph has a vertex cover (for example if Q contains all the vertices in the graph), yet we want to optimize the vertex cover; that is, find a minimum vertex cover. If every edge has an endpoint to one of the vertices in a vertex cover, then at most one matched edge can be incident to any single vertex in the cover. This result, stated in the theorem below, was first published in 1931 by the Hungarian mathematicians Dénes König and (independently) Jenö Egerváry, and as noted above was the inspiration behind the Augmenting Path Algorithm.

Theorem 5.11 (König-Egerváry Theorem) For a bipartite graph G, the size of a maximum matching of G equals the size of a minimum vertex cover for G.

This theorem again provides a basis for the optimization question. Consider the graph G_4 from page 157 (reproduced below) in the discussion leading to Hall's Marriage Theorem. Based on the König-Egerváry Theorem, to show the matching in blue is maximum we need to find a vertex cover of size 3. One such cover is shown below. You should check that every edge has an endpoint that is either d, g or h.

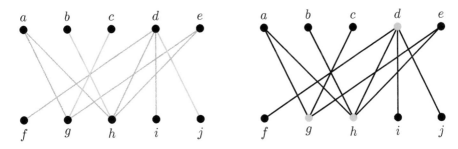

In most cases, a minimum vertex cover for a bipartite graph will require some vertices from both pieces of the vertex partition. It should not come to much surprise that the Augmenting Path Algorithm can be used to find a vertex cover. In fact, many texts include the procedure below as part of the algorithm itself.

Vertex Cover Method

1. Let $G = (X \cup Y, E)$ be a bipartite graph.

2. Apply the Augmenting Path Algorithm and mark the vertices considered throughout its final implementation.

3. Define a vertex cover Q as the unmarked vertices from X and the marked vertices from Y.

4. Q is a minimum vertex cover for G.

In Step 2, a vertex is marked if it was considered during the final step in the implementation of the Augmenting Path Algorithm. Note that this is not just the vertices in U, the unsaturated vertices from X, but also any vertex that was reached through an alternating path that originated at a vertex from U. Thus the unmarked vertices will be those that are never mentioned during the final step of the implementation of the Augmenting Path Algorithm.

Example 5.6 Apply the Vertex Cover Method to the output graph from Example 5.5.

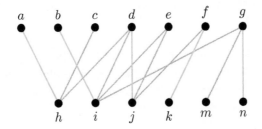

Solution: Recording the vertices considered throughout the last step of the Augmenting Path Algorithm, the marked vertices from X are a, b, c, d, and e, and the marked vertices from Y are h, i, and j. This produces the vertex cover $Q = \{f, g, h, i, j\}$ of size 5, shown below. Recall that the maximum matching contained 5 edges.

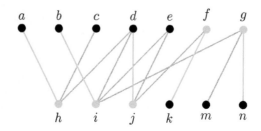

Note that more than one minimum vertex cover may exist for the same graph, just as more than one maximum matching may exist. In the example above, we found one such vertex cover through the matching found using the Augmenting Path Algorithm, though the set $Q' = \{g, h, i, j, k\}$ is also a valid minimum vertex cover.

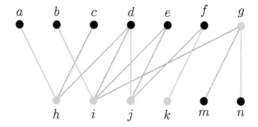

By finding a minimum vertex cover, we are able to answer the optimization question: how large of a matching can be formed. We conclude this section with one more application of bipartite graph matchings, called distinct representatives, before moving to matchings on other types of graphs.

Definition 5.12 Given a collection of finite nonempty sets $S_1, S_2, \ldots, S_n$ (where $n \geq 1$), a system of **distinct representatives** is a collection $r_1, r_2, \ldots, r_n$ so that r_i is a member of set S_i and $r_i \neq r_j$ for all $i \neq j$ (for all $i, j = 1, 2, \ldots, n$).

In less technical terms, the idea of distinct representatives is that a collection of groups each need their own representative and no two groups can have the same representative.

Example 5.7 During faculty meetings at a small liberal arts college, multiple committees provide a report to the faculty at large. These committees often overlap in membership, so it is important that, for any given year, a person is not providing the report for more than one committee. Find a collection of distinct representatives for the groups listed below.

Committee	Members		
Admissions Council	Ivan	Leah	Sarah
Curriculum Committee	Kyle	Leah	
Development and Grants	Ivan	Kyle	Norah
Honors Program Council	Norah	Sarah	Victor
Personnel Committee	Sarah	Victor	

Solution: Begin by forming a bipartite graph where X consists of the committees and Y the faculty members, and draw an edge between two vertices if a person is a member of that committee.

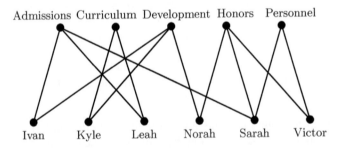

A collection of distinct representatives is modeled as a matching. Note that we are interested in each committee having a representative, not every person being a representative. Thus we want an X-matching. One such matching is shown below.

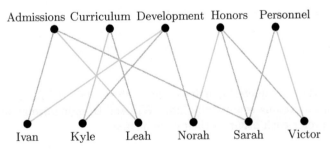

5.3   Stable Marriages

Up to this point, we have only been concerned with finding the largest matching possible; but in many circumstances, one pairing may be preferable over another. When taking preferences into account, we no longer focus on whether two items can be paired but rather which pairing is best for the system. Initially, we will only consider those situations that can be modeled by a bipartite graph with an equal number of items in X and Y. In addition, previous models demonstrated undesirable pairings by leaving off the edge in the graph; but with preferences added, we include every edge possible in the bipartite graph. This means the underlying graph will be a complete bipartite graph.

The preference model for matchings is often referred to as the *Stable Marriage Problem* to parallel Hall's Marriage Theorem, and our terminology will reflect the marriage model. We start with two distinct, yet equal sized, groups of people, usually men and women, who have ranked the members of the other group. The stability of a matching is based on if switching two matched edges would result in happier couples.

Definition 5.13 A perfect matching is **stable** if no unmatched pair is **unstable**; that is, if x and y are not matched but both rank the other higher than their current partner, then x and y form an unstable pair.

In essence, when pairing couples into marriages we want to ensure no one will leave their current partner for someone else. To better understand stability, consider the following example.

Example 5.8 Four men and four women are being paired into marriages. Each person has ranked the members of the opposite sex as shown below. Draw a bipartite graph and highlight the matching Anne–Rob, Brenda–Ted, Carol–Stan, and Diana–Will. Determine if this matching is stable. If not, find a stable matching and explain why no unstable pair exists.

Anne: t > r > s > w	Rob: a > b > c > d
Brenda: s > w > r > t	Stan: a > c > b > d
Carol: w > r > s > t	Ted: c > d > a > b
Diana: r > s > t > w	Will: c > b > a > d

Solution: The complete bipartite graph is shown below with the matching in blue. This matching is not stable since Will and Brenda form an unstable pair, as they prefer each other to their current mate.

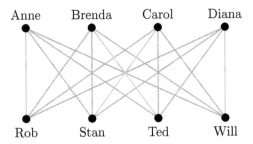

Switching the unstable pairs produces the matching below (Anne – Rob, Brenda – Will, Carol – Stan and Diana – Ted).

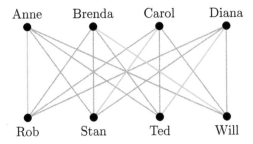

Note that this matching is not stable either since Will and Carol prefer each other to their current mate. Switching the pairs produces a new matching (Anne – Rob, Brenda – Stan, Carol – Will, Diana – Ted) shown below.

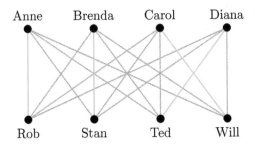

This final matching is in fact stable due to the following: Will and Carol are paired, and each other's first choice; Anne only prefers Ted over Rob, but Ted does not prefer Anne over Diana; Brenda does not prefer anyone over Stan; Diana prefers either Rob or Stan over Ted, but neither of them prefers Diana to their current mate.

While searching for an unstable pair and switching the matching may eventually result in a stable matching (in fact, it will always eventually land on a stable matching), a better procedure exists for finding a stable matching. The algorithm below is named for David Gale and Lloyd Shapley, the two American mathematicians and economists who published this algorithm in 1962 [17]. Though the original paper discussed the stable marriage problem, it was mainly concerned with college admissions where the size of the applicant pool differs from the number of colleges and a college accepts more than one student. Their brilliant approach has been modified to work on problems that cannot be modeled by a complete bipartite graph, a few of which appear in the next section. In addition, their work led to further studies on economic markets, one of which awarded Shapley (along with his collaborator Alvin Roth) the 2012 Nobel Prize in Economics.

Gale-Shapley Algorithm

Input: Preference rankings of n women and n men.

Steps:

1. Each man proposes to the highest ranking woman on his list.

2. If every woman receives only one proposal, this matching is stable. Otherwise move to Step (3).

3. If a woman receives more than one proposal, she

   (a) accepts if it is from the man she prefers above all other currently available men and rejects the rest; or,

   (b) delays with a maybe to the highest ranked proposal and rejects the rest.

4. Each man now proposes to the highest ranking unmatched woman on his list who has not rejected him.

5. Repeat Steps (2) – (4) until all people have been paired.

Output: Stable Matching.

The Gale-Shapley Algorithm is written so that the men are always proposing, giving the women the choice to accept, reject, or delay. This produces an asymmetric algorithm, meaning a different outcome could occur if the women were proposing, which is demonstrated in the next two examples.

Example 5.9 Apply the Gale-Shapley Algorithm to the rankings from Example 5.8, which are reproduced below.

Anne:	t	> r	> s	> w	Rob:	a	> b	> c	> d
Brenda:	s	> w	> r	> t	Stan:	a	> c	> b	> d
Carol:	w	> r	> s	> t	Ted:	c	> d	> a	> b
Diana:	r	> s	> t	> w	Will:	c	> b	> a	> d

Solution: For record keeping, in each round of proposals we indicate if a proposal is accepted by a check (✓), a rejection by an X and a delay as **?**.

Step 1: The initial proposals are Rob – Anne, Stan – Anne, Ted – Carol, and Will – Carol. Note that we think of these proposals as simultaneous and a woman does not make her decision until all proposals are made for a given round.

Step 2: Since not all the proposals are different, Anne and Carol need to make some decisions. First, neither Rob nor Stan are Anne's top choice so she rejects the lower ranked one (Stan) and says maybe to the other (Rob). Next, Will is Carol's top choice so she accepts his proposal and rejects Ted.

Rob	Anne	?
Stan	Anne	X
Ted	Carol	X
Will	Carol	✓

Step 3: The remaining men (Rob, Stan and Ted) propose to the next available woman on their preference list. Rob proposes again to Anne since she delayed in the last round. Stan proposes to Brenda; he cannot propose to Anne since she rejected him previously and cannot propose to Carol since she is already paired. Ted proposes to Diana.

Step 4: Since all the proposals are different (no woman received more than one proposal), the women must all accept.

Rob	Anne	✓
Stan	Brenda	✓
Ted	Diana	✓
Will	Carol	✓

Output: The matching shown above is stable.

As noted above, the Gale-Shapley Algorithm is asymmetric and in fact favors the group making the proposals. In the example above, two men are paired with their top choice, one with his second, and one with his third. Even though the same holds for the women (two first choices, one second, and one third), if the women were the ones proposing we would likely see an improvement in their overall happiness. This is outlined in the example below.

Example 5.10 Apply the Gale-Shapley Algorithm to the rankings from Example 5.8 with the women proposing.

Solution:

Step 1: The women all propose to their top choice. The initial proposals are Anne – Ted, Brenda – Stan, Carol – Will, and Diana – Rob.

Step 2: Since all the proposals are different (no man received more than one proposal), the men must all accept.

Anne	Ted	✓
Brenda	Stan	✓
Carol	Will	✓
Diana	Rob	✓

Output: The matching shown above is stable.

The examples above demonstrate two important properties of stable matching. First, the group proposing in the Gale-Shapley Algorithm is more likely to be happy. This is especially true if the top choices are all different for the proposers, as happened above. In Example 5.10, the men were required to accept the proposal they received since all the proposal were different. This ensured the women were all paired with their first choice, whereas only one man was paired with his first choice, two with their third choice, and one with his fourth choice. Though this may seem more imbalanced than the matching in Example 5.9, it is still a stable matching, which demonstrates the second item about stable matchings. There is no guarantee that a unique stable matching exists. In fact, many examples have more than one stable matching possible. The important concept to remember is that for a complete bipartite graph with rankings, a stable matching will *always* exist. If we generalize this to other types of graphs, the same may not hold.

One last note on the history of this procedure. Though the algorithm is correctly attributed to Gale and Shapley, it had been implemented about a decade earlier in the pairing of hospitals and residents. The residency selection process was poorly managed prior to the foundation of the National

Resident Matching Program in 1952 by medical students. In 1984 Alvin Roth proved that the algorithm used by the NRMP was a modification of the Gale-Shapley Algorithm. Its implementation had the hospitals "proposing" to the medical students, which meant that the residency programs were favored over the applicants. The algorithm was readjusted in 1995 to have the applicants proposing to the residency programs, ensuring the applicants' preferences are favored. The NRMP today encompasses more than 40,000 applicants and 30,000 positions.[37]

There has been extensive study into variations on the Stable Marriage Problem described above. We will discuss two of these: Unacceptable Partners and Stable Roommates. We will begin with the Unacceptable Partners problem since it still has a bipartite graph as its underlying structure. The Stable Roommates problem will be studied in the next section dealing with matchings on graphs that are not bipartite. Further generalizations to the Stable Marriage Problem can be found in Exercise 16, [20] and [21].

Unacceptable Partners

Look back at the preferences in Example 5.8. By having each person rank all others of the opposite sex, we assume that all of these potential matches are acceptable. This is very clearly not accurate to a real world scenario — some people should never be married if even they are the only pair left. To adjust for this, we introduce the notion of an *unacceptable partner*. Consider the rankings below, where if a person is missing from the ranking, then they are deemed unacceptable (so Will is unacceptable to Diana and only Anne and Brenda are acceptable to Stan).

Anne: $t > r > s > w$	Rob: $a > b > c > d$	
Brenda: $w > r > t$	Stan: $a > b$	
Carol: $w > r > s > t$	Ted: $c > d > a > b$	
Diana: $s > r > t$	Will: $c > b > a$	

We are still looking for a matching in a bipartite graph, only now the graph is not complete. We must adjust our notion of a stable matching, since it is possible that not all people could be matched (think of a confirmed bachelor; he would label all women as unacceptable). Under these new conditions, a matching (with unacceptable partners) is *stable* so long as no unmatched pair x and y exist such that x and y are both acceptable to each other, and each is either single or prefers the other to their current partner. To account for this new definition of stable, we must make two minor adjustments to the Gale-Shapley Algorithm.

Gale-Shapley Algorithm (with Unacceptable Partners)

Input: Preference rankings of n women and n men.

Steps:

1. Each man proposes to the highest ranking woman on his list.

2. If every woman receives only one proposal from someone they deem acceptable, they all accept and this matching is stable. Otherwise move to Step (3).

3. If the proposals are not all different, then each woman:

   (a) rejects a proposal if it is from an unacceptable man;

   (b) accepts if the proposal is from the man she prefers above all other currently available men and rejects the rest; or

   (c) delays with a maybe to the highest ranked proposal and rejects the rest.

4. Each man now proposes to the highest ranking unmatched woman on their list who has not rejected him.

5. Repeat Steps (2) – (4) until all people have been paired or until no unmatched man has any acceptable partners remaining.

Output: Stable Matching.

The major change here is in dealing with unacceptable partners. As with the original form of the Gale-Shapley Algorithm, this new version always produces a stable matching. In addition, the algorithm is written so that the men are proposing, but as before, this can be modified so that the women are proposing.

Example 5.11 Apply the Gale-Shapley Algorithm to the rankings on page 173 to find a stable matching.

Solution:

Step 1: The initial proposals are Rob – Anne, Stan – Anne, Ted – Carol, and Will – Carol.

Step 2: Since not all the proposals are different, Anne and Carol need to make some decisions. First, since Stan is unacceptable to Anne she rejects him and since Rob is not her top choice she says maybe. Next, Will is Carol's top choice so she accepts his proposal and rejects Ted.

Rob	Anne	?
Stan	Anne	X
Ted	Carol	X
Will	Carol	✓

Step 3: The remaining men propose to the next available woman on their preference list. Rob proposes again to Anne since she delayed in the last round. Stan proposes to Brenda; he cannot propose to Anne since she rejected him previously. Ted proposes to Diana.

Step 4: Even though all proposals are different, Brenda rejects Stan since he is an unacceptable partner. The other two women say maybe since their proposals are not from their top choice.

Rob	Anne	?
Stan	Brenda	X
Ted	Diana	?
Will	Carol	✓

Step 5: Rob proposes again to Anne and Ted proposes to Diana since the women said maybe in the last round. Stan does not have any acceptable partners left, and so must remain single.

Step 6: Anne and Diana accept their proposals since all proposals are different.

Rob	Anne	✓
Stan		
Ted	Diana	✓
Will	Carol	✓

Output: The matching shown above is stable. Note that Stan and Brenda remain unmatched.

As the example above illustrates, in a scenario with unacceptable partners a stable matching can exist with not all people paired. The version of this example where the women propose appears in Exercise 5.14. A modification of this procedure to allow for situations with an unequal number of men and women appears in Exercise 5.16.

5.4   Matchings in Non-Bipartite Graphs

Up to this point we have only discussed matchings inside of bipartite graphs. Although these model many problems that are solved using a matching, there are some problems that are best modeled with a graph that is not bipartite. Consider the following scenario:

> Bruce, Evan, Garry, Hank, Manny, Nick, Peter, and Raj decide to go on a week-long canoe trip in Guatemala. They must divide themselves into pairs, one pair for each of four canoes, where everyone is only willing to share a canoe with a few of the other travelers.

Modeling this as a graph cannot result in a bipartite graph since there are not two distinct groups that need to be paired, but rather one large group that must be split into pairs.

Example 5.12   The group of eight men from above have listed who they are willing to share a canoe with. This information is shown below in the table, where a Y indicates a possible pair. Note that these relationships are symmetric, so if Bruce will share a canoe with Manny, then Manny is also willing to share a canoe with Bruce.

	Bruce	Evan	Garry	Hank	Manny	Nick	Peter	Raj
Bruce	·	·	·	·	Y	Y	·	Y
Evan	·	·	·	Y	Y	Y	·	·
Garry	·	·	·	Y	·	·	Y	Y
Hank	·	Y	Y	·	·	·	Y	·
Manny	Y	Y	·	·	·	·	·	Y
Nick	Y	Y	·	·	·	·	·	Y
Peter	·	·	Y	Y	·	·	·	·
Raj	Y	·	Y	·	Y	Y	·	·

Model this information as a graph. Find a perfect matching or explain why no such matching exists.

Solution: The graph is shown below where an edge represents a potential pairing into a canoe.

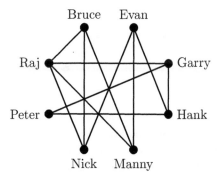

Note that Peter can only be paired with either Garry or Hank. If we choose to pair Peter and Garry, then Hank must be paired with Evan, leaving Nick and Manny to each be paired with one of Raj and Bruce. One possible matching is shown below. Since all people have been paired, we have a perfect matching.

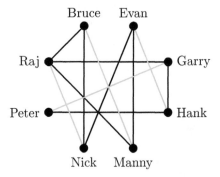

Finding matchings in general graphs is often more complex than in bipartite graphs, in part due to the fact that only some of the results from Section 5.2 still apply. In particular, Berge's Theorem holds (M is maximum if and only if G has no M-augmenting paths) yet the Augmenting Path Algorithm only applies to bipartite graphs. The main sticking point is that in finding alternating paths from an unsaturated vertex, there could be more than one path to x and only investigating one would miss an augmenting path. For example, if we are searching for alternating paths from u to x in the graph below, we might choose $u\,a\,b\,x$. If instead we chose $u\,a\,b\,c\,x$, we could continue the alternating path to find an augmenting path to y (namely, $u\,a\,b\,c\,x\,y$).

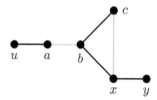

Jack Edmonds devised a modification to the Augmenting Path Algorithm that works on general graphs, making use of these possible odd cycles. This algorithm is more powerful than we need for small examples, as these can usually be determined through inspection and some reasoning. For further information on Edmonds' Blossom Algorithm, see [40].

Example 5.13 Halfway through the canoe trip from Example 5.12, Raj will no longer share a canoe with Garry, and Hank angered Evan so they cannot share a canoe. Update the graph model and determine if it is now possible to pair the eight men into four canoes.

Solution: The updated graph only removes two edges, but in doing so the graph is now disconnected. A disconnected graph could have a perfect matching so long as each component itself has a perfect matching. In this case, a perfect matching is impossible since Peter, Garry, and Hank form one component and so at most two of them can be paired. Likewise, the other component contains five people and so at most four can be paired. Thus there is no way to pair the eight men into four canoes.

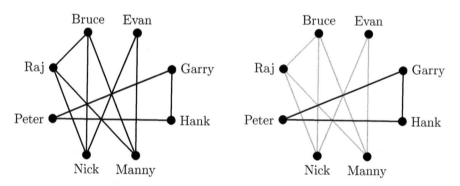

Stable Roommates

The Stable Roommate Problem is a modification of the Stable Marriage problem, only now the underlying graph is not bipartite. Each person ranks the others and we want a stable matching; that is, a matching so that two unpaired people do not both prefer each other to their current partner. Similar to the Augmenting Path Algorithm being unavailable for use on general graphs, the

Gale-Shapley Algorithm does not work on the Stable Roommate Problem since it is based on two distinct groups ranking members of the other group. This first example finds all possible pairings and examines if they are stable.

Example 5.14 Four women are to be paired as roommates. Each woman has ranked the other three as shown below. Find all possible pairings and determine if any are stable.

$$\begin{array}{rcccc} \text{Emma:} & l & > & m & > & z \\ \text{Leena:} & m & > & e & > & z \\ \text{Maggie:} & e & > & z & > & l \\ \text{Zara:} & e & > & l & > & m \end{array}$$

Solution: There are three possible pairings, only one of which is stable.

- Emma – Leena and Maggie – Zara
  This is stable since Emma is with her first choice and the only person Leena prefers over Emma is Maggie, but Maggie prefers Zara over Leena.

- Emma – Maggie and Leena – Zara
  This is not stable since Emma prefers Leena over Maggie and Leena prefers Emma over Zara.

- Emma – Zara and Leena – Maggie
  This is not stable since Emma prefers Maggie over Zara and Maggie prefers Emma over Leena.

Recall the Gale-Shapley Algorithm showed that a stable matching on a bipartite graph (where the partition sets have equal size) will always exist, yet it is possible in the Stable Roommate Problem that a stable matching will fail to exist.

Example 5.15 Before the four women from Example 5.14 are paired as roommates, Maggie and Zara get into an argument, causing them to adjust their preference lists. Determine if a stable matching exists.

$$\begin{array}{rcccc} \text{Emma:} & l & > & m & > & z \\ \text{Leena:} & m & > & e & > & z \\ \text{Maggie:} & e & > & l & > & z \\ \text{Zara:} & e & > & l & > & m \end{array}$$

Solution: There are three possible pairings, none of which are stable.

- Emma – Leena and Maggie – Zara
  This is not stable since Leena prefers Maggie over Emma and Maggie prefers Leena over Zara.

- Emma – Maggie and Leena – Zara
  This is not stable since Emma prefers Leena over Maggie and Leena prefers Emma over Zara.

- Emma – Zara and Leena – Maggie
  This is not stable since Emma prefers Maggie over Zara and Maggie prefers Emma over Leena.

The previous two examples used an exhaustive method to find a stable matching or to determine if no such matching exists. While this is not too difficult with a small number of vertices, it becomes computationally impractical as the number of vertices grows. Recall from Section 3.1 that the number of ways to pair n items, when n is even, is $(n-1)!!$. Thus for a Stable Roommate Problem, we would need to check 15 pairings when there are 6 people and check 105 for 8 people (such as the canoe example). An efficient algorithm for the Stable Roommate Problem was first published in 1985 by the computer scientist Robert Irving; for further information, see Exercise 5.17, or [20].

5.5 Exercises

5.1 Draw the complete bipartite graphs $K_{2,3}$, $K_{1,4}$, and $K_{3,5}$.

5.2 Find a maximum matching for the graph from Example 1.4 on page 6.

5.3 Below is a graph with a matching M shown in blue.

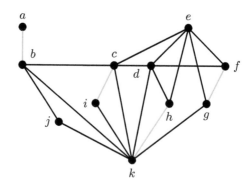

(a) Find an alternating path starting at a. Is this path augmenting?

(b) Find an augmenting path in the graph or explain why none exists.

(c) Is M a maximum matching? maximal matching? perfect matching? Explain your answer. If M is not maximum, find a matching that is maximum.

5.4 Each of the graphs below has a matching shown in blue. Complete the following steps for both:

(i) Find an alternating path starting at vertex a.

(ii) Is this path augmenting? Explain your answer.

(iii) Use the Augmenting Path Algorithm to find a maximum matching.

(iv) Use the Vertex Cover Method to find a minimum vertex cover.

(a)

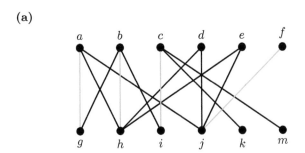

(b)

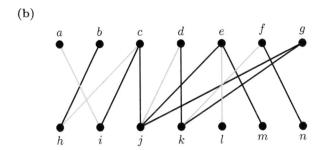

5.5 Find a maximum matching for each of the graphs below. Include an explanation as to why the matching is maximum.

(a)

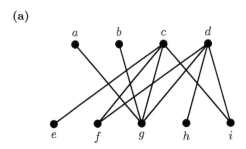

(b)

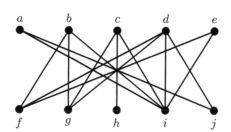

(c)

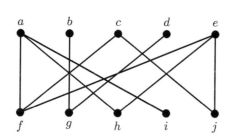

(d)

(e)

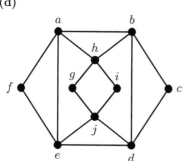
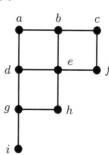

(f)

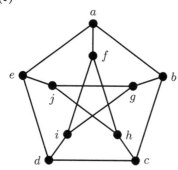

5.6 Using the graphs from Exercise 5.5,

(a) Determine which graphs are bipartite.

(b) For each of the graphs that are bipartite, find a minimum vertex cover. Verify that the size of the matching found in Exercise 5.5 equals the size of your vertex cover.

5.7 The Roanoke Ultimate Frisbee League is organizing a Contra Dance. The fifteen members must be split into male-female pairs, though not all people are willing to dance with each other. The graph below models those who can be paired (as both people find the other acceptable). Find a maximum matching and explain why a larger matching does not exist.

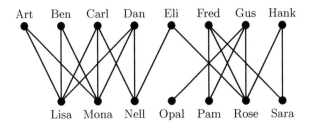

5.8 Seven committees must elect a chairperson to represent them at the end-of-year board meeting; however, some people serve on more than one committee and so cannot be elected chairperson for more than one committee. Based on the membership lists below, determine a system of distinct representatives for the board meeting.

Committee	Members			
Benefits	Agatha	Dinah	Evan	Vlad
Computing	Evan	Nancy	Leah	Omar
Purchasing	George	Vlad	Leah	
Recruitment	Dinah	Omar	Agatha	
Refreshments	Nancy	George		
Social Media	Evan	Leah	Vlad	Omar
Travel Expenses	Agatha	Vlad	George	

5.9 Each year, the chair of the mathematics department must determine course assignments for the faculty. Each professor has submitted a list of the courses he or she wants to teach. Find a system of assignments where each professor will teach exactly one of the remaining courses or explain why none exists.

Professor	Preferred Courses			
Dave	Abstract Algebra	Real Analysis	Statistics	Calculus
Roland	Statistics	Geometry	Calculus	
Chris	Calculus	Geometry		
Adam	Statistics	Calculus		
Hannah	Abstract Algebra	Real Analysis	Topology	
Maggie	Abstract Algebra	Real Analysis	Geometry	Topology

5.10 Instead of pairing a professor with only one course of their preference from Exercise 9, now the mathematics department chair must pair each professor with two of the courses from their (expanded) list.

(a) Describe how to turn this into a matching problem where a solution is given in terms of a perfect matching.

(b) Find a perfect matching for the professors and their preferred course list shown below or explain why none exists.

Professor	Preferred Courses		
Dave	Abstract Algebra	Real Analysis	Number Theory
	Calculus II	Calculus I	Statistics
Roland	Vector Calculus	Discrete Math	Statistics
	Calculus II	Geometry	Calculus I
Chris	Vector Calculus	Real Analysis	Discrete Math
	Statistics	Geometry	Calculus I
Adam	Statistics	Calculus I	Number Theory
	Geometry	Differential Equations	
Hannah	Abstract Algebra	Real Analysis	Number Theory
	Linear Algebra	Topology	
Maggie	Abstract Algebra	Real Analysis	Linear Algebra
	Geometry	Topology	Calculus II

5.11 The students in a geometry course are paired each week to present homework solutions to the class. In the table below, a possible pair is indicated by a Y. Find a way to pair the students or explain why none exists.

	Al	Brie	Cam	Fred	Hans	Megan	Nina	Rami	Sal	Tina
Al	.	Y	.	.	Y	.	.	Y	.	Y
Brie	Y	.	Y	Y	Y	Y	.	Y	Y	.
Cam	.	Y	.	.	Y	.	Y	Y	.	.
Fred	.	Y	.	.	.	Y	Y	.	.	Y
Hans	Y	Y	Y	.	.	.	.	Y	Y	Y
Megan	.	Y	.	Y	.	.	Y	Y	Y	.
Nina	.	.	Y	Y	.	Y	.	.	Y	.
Rami	Y	Y	Y	.	Y	Y	.	.	Y	.
Sal	.	Y	.	.	Y	Y	Y	Y	.	.
Tina	Y	.	.	Y	Y	.	.	.	.	.

5.12 Apply the Gale-Shapley Algorithm to the set of preferences below with

(a) the men proposing

(b) the women proposing

Alice:	r	>	s	>	t	>	v	Rich:	a	> d > b > c		
Beth:	s	>	r	>	v	>	t	Stefan:	a	> c > d > b		
Cindy:	v	>	t	>	r	>	s	Tom:	c	> b > d > a		
Dahlia:	t	>	v	>	s	>	r	Victor:	c	> d > b > a		

5.13 Apply the Gale-Shapley Algorithm to the set of preferences below with

(a) the men proposing

(b) the women proposing

Edith: $l > n > o > m > p$ Liam: $f > e > h > g > i$

Faye: $n > l > m > o > p$ Malik: $e > i > g > f > h$

Grace: $p > m > o > n > l$ Nate: $f > g > i > h > e$

Hanna: $p > n > o > l > m$ Olaf: $i > e > f > g > h$

Iris: $p > o > m > n > l$ Pablo: $f > h > g > e > i$

5.14 Apply the Gale-Shapley Algorithm (with Unacceptable Partners) to the preferences from Example 5.11 with the women proposing.

5.15 Apply the Gale-Shapley Algorithm (with Unacceptable Partners) to the preferences below with

(a) the men proposing

(b) the women proposing

Edith: $l > n > m$ Liam: $f > e > h > g$

Faye: $n > l > m > o > p$ Malik: $e > h > i > f$

Grace: $m > o > n > l$ Nate: $g > f > i$

Hanna: $p > o > l > m$ Olaf: $i > e > f$

Iris: $p > m > n > l$ Pablo: $f > h > g > i$

5.16 In each of the examples where the Gale-Shapley Algorithm is utilized, we have required that the number of men equals the number of women. Just as we were able to modify the algorithm for instances where some people are deemed unacceptable, we can modify the algorithm to account for unequal numbers. To do this, we introduce ghost participants in order to equalize the gender groups. These ghosts are deemed unacceptable by those of the opposite sex, and in turn find no person of the opposite sex acceptable. Using this modification, find a stable set of marriages for the preferences listed below.

Alice:	p > r > s > t				Peter:	b > a > c > d > e				
Beth:	r > p > s > t				Rich:	c > b > e > d > a				
Carol:	t > p > s > r				Saul:	a > b > c > d > e				
Diana:	t > s > r > p				Teddy:	e > c > d > a > b				
Edith:	r > s > t > p									

Projects

5.17 Section 5.4 described the challenges that arise in the Stable Roommate Problem that do not occur in the Stable Marriage Problem. Although one was not introduced, algorithms do exist for the Stable Roommate Problem that either find a stable matching or explain why none exists. Research one of these methods and apply it to the preferences shown below. (See [20] for more information.)

Bruce:	m > n > r > g > e > p > h
Evan:	h > n > m > p > b > r > g
Garry:	p > r > h > m > b > e > n
Hank:	g > e > p > n > r > m > b
Manny:	r > e > b > g > h > p > n
Nick:	r > b > e > m > p > h > g
Peter:	h > g > r > n > e > b > m
Raj:	b > m > g > n > e > h > p

5.18 In Section 4.4, we discussed an approximation algorithm for the metric Traveling Salesman Problem that made use of a minimum spanning tree. Christofides' Algorithm, outlined below, uses the same basic backbone of using a minimum spanning tree, but also makes use of a matching within the graph. Christofides' Algorithm, first introduced in 1976, had the best performance ratio of any approximation algorithm until a modified version was published in 2015.

Apply Christofides' Algorithm to the graph from Example 4.12 and Exercise 4.12.

Christofides' Algorithm

Input: Weighted complete graph K_n, where the weight function w satisfies the triangle inequality.

Steps:

1. Find a minimum spanning tree T for K_n.

2. Let X be the set of all vertices of odd degree in T.

3. Create a weighted complete graph H whose vertex set is X and where the weight of the edges is taken from the weights given in K_n.

4. Find a perfect matching M on H of minimum weight.

5. Create a graph G by adding the matched edges from M to the tree T obtained in Step (1).

6. Find an Euler circuit for G.

7. Convert the Eulerian circuit into a Hamiltonian cycle by skipping any previously visited vertex (except for the starting and ending vertex).

8. Calculate the total weight.

Output: Hamiltonian cycle for K_n.

Chapter 6

Graph Coloring

In the previous chapter we discussed the application of graph matching to a problem where items from two distinct groups must be paired. An important aspect of this pairing is that no item could be paired more than once. Compare that with the following scenario:

> Five student groups are meeting on Saturday, with varying time requirements. The staff at the Campus Center need to determine how to place the groups into rooms while using the fewest rooms possible.

Although we can think of this problem as pairing groups with rooms, there is no restriction that a room can only be used once. In fact, to minimize the number of rooms used, we would hope to use a room as often as possible. This chapter explores graph coloring, a strategy often used to model resource restrictions. But before we get into the heart of this graph model, we begin with a historically significant problem, known as the Four Color Theorem.

6.1 Four Color Theorem

In 1852 Augustus De Morgan sent a letter to his colleague Sir William Hamilton (the same mathematician who introduced what we now call Hamiltonian cycles) regarding a puzzle presented by one of his students, Frederick Gutherie (though Gutherie later clarified that the question originated from his brother, Francis). This question was known for over a century as the *Four Color Conjecture*, and can be stated as

> Any map split into contiguous regions can be colored using at most four colors so that no two bordering regions are given the same color.

An important aspect of this conjecture is that a region, such as a country or state, cannot be split into two disconnected pieces. For example, the state of Michigan is split into the Lower Peninsula and the Upper Peninsula and so is not a contiguous region; thus the contiguous United States does not satisfy the hypothesis of the Four Color Conjecture. However, it is still possible to color the lower 48 states using 4 colors (try it!).

The Four Color Conjecture started as a map coloring problem, yet migrated into a graph coloring problem. In the late 19th century, Alfred Kempe studied the dual problem where each region on a map was represented by a vertex and an edge exists between two vertices if their corresponding regions share a border. This approach was extensively used in the mid-20th century as the study of graph theory exploded with the advent of the computer. The search for a proper map coloring is now reduced to a proper vertex coloring (more commonly referred to as just a coloring) for a *planar graph*. A graph is planar if it can be drawn so that no edges cross. For more information on planar graphs see Section 7.6.

Definition 6.1 A proper **k-coloring** of a graph G is an assignment of colors to the vertices of G so that no two adjacent vertices are given the same color and exactly k colors are used.

Most problems on graph coloring are optimization problems since we want to minimize the number of colors used; that is, find the lowest value of k so that G has a proper k-coloring. The example below demonstrates how to convert a map into a graph and how to convert a graph coloring back into a coloring of a map. Note that beyond small examples, we rarely use color names (red, blue, green, etc.) but rather refer to color numbers (color 1, 2, 3, etc.) since names of colors get more complicated as we move beyond the standard 6 to 10 colors.

Example 6.1 Find a coloring of the map of the counties of Vermont and explain why three colors will not suffice.

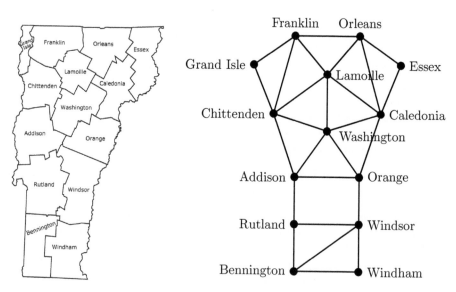

Solution: First note that each county is given a vertex and two vertices are adjacent in the graph when their respective counties share a border. One possible coloring is shown below. Note that Lamoille County is surrounded by five other counties. If we try to alternate colors amongst these five counties, for example Orleans – 1, Franklin – 2, Chittenden – 1, Washington – 2, we still need a third color for the fifth county (Caledonia – 3). Since Lamoille touches each of these counties, we know it needs a fourth color.

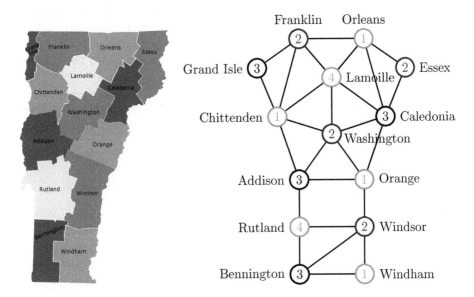

The Four Color Conjecture intrigued mathematicians in part due to its simplicity but also because of the numerous false proofs. Some of the most brilliant mathematicians of the 19th and 20th centuries incorrectly believed they had proven the conjecture and it was not until 1976 that a correct proof was published by Kenneth Appel and Wolfgang Haken. Their proof was not widely accepted for some time both for its use of a computer program (the first major theorem in mathematics to do so) and the difficulty in checking their work. The proof filled over 900 pages, contained over 10,000 diagrams, took thousands of hours of computational time, and when printed stood about four feet in height. Mathematicians have since offered refinements to the Appel-Haken argument; in particular, Niel Robertson, Daniel Sanders, Paul Seymour and Robin Thomas published an updated version in 1994 that not only reduced the number of pages in the proof, but also made it feasible for others to verify their results on a standard home computer. For further history of the Four Color Theorem, see [41].

6.2   Coloring Bounds

For the remainder of this chapter, we will explore graph colorings for graphs that may or may not be planar, mainly since we already know that planar graphs need at most 4 colors and so there is not much room for further exploration. Any graph we consider can be simple or have multi-edges but cannot have loops, since a vertex with a loop could never be assigned a color. In any graph coloring problem, we often tackle the existence, construction, and optimization question at once; that is, we want to determine the smallest value for k for which a graph has a k-coloring. This value for k is called the *chromatic number* of a graph.

Definition 6.2 The ***chromatic number*** $\chi(G)$ of a graph is the smallest value k for which G has a proper k-coloring.

In order to determine the chromatic number of a graph, we often need to complete the following two steps:

(1) Find a coloring of G using k colors.

(2) Show why fewer colors will not suffice.

At times it can be quite complex to show a graph cannot be colored with fewer colors. There are a few properties of graphs and the existence of certain subgraphs that can immediately provide a basis for these arguments.

Look back at Example 6.1 about coloring the counties in Vermont and the discussion of alternating colors around a central vertex. In doing so, we were using one of the most basic properties in graph coloring: the number of colors needed to color a cycle. Recall that a cycle on n vertices is denoted C_n. The examples below show optimal colorings of C_3, C_4, C_5, and C_6.

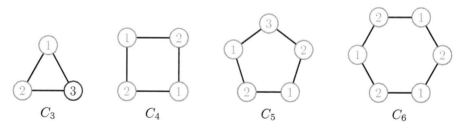

Notice that in all the graphs we try to alternate colors around the cycle. When n is even, we can color C_n in two colors since this alternating pattern can be completed around the cycle. When n is odd, we need three colors for C_n since the final vertex visited when traveling around the cycle will be adjacent to a vertex of color 1 and of color 2. This was demonstrated in the coloring of the five counties surrounding Lamoille County in Example 6.1.

The next structure that provides additional reasoning for the lower bound of the chromatic number is based upon an odd cycle. Again, referencing the coloring of the counties in Vermont, we used the odd cycle around Lamoille County to explain why at least 3 colors were needed. However, we showed that in fact 4 colors were required since the Lamoille vertex is adjacent to each of vertices in the surrounding odd cycle. This structure is often referred to as a wheel.

Definition 6.3 A *wheel* W_n is a graph in which n vertices form a cycle around a central vertex that is adjacent to each of the vertices in the cycle.

The first few wheels are shown below. Note that when n is odd, we get a scenario similar to that of Lamoille county from Example 6.1, and thus requiring 4 colors. In general, we can use odd wheels to explain why 3 colors will not suffice.

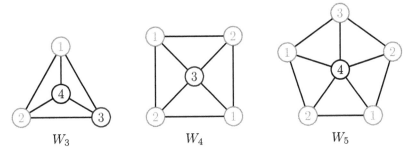

$$W_3 \qquad W_4 \qquad W_5$$

The final structure we search for within a graph is based on the notion of a complete graph. Recall that in complete graphs each vertex is adjacent to every other vertex in the graph. Thus if we assign colors to the vertices, we cannot use a color more than once. Possible colorings of a few complete graphs are shown below.

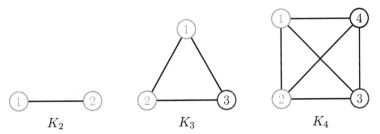

$$K_2 \qquad K_3 \qquad K_4$$

When a complete graph appears as a subgraph within a larger graph, we call it a *clique*.

Definition 6.4 A *clique* in a graph is a subgraph that is itself a complete graph. The *clique size* of a graph G, denoted $\omega(G)$, is the largest value of n for which G contains K_n as a subgraph.

Knowing the clique size of a graph automatically provides a nice lower bound for the chromatic number. For example, if G contains K_5 as a subgraph, then we know this portion of the graph needs at least 5 colors. Thus $\chi(G) \geq 5$. Thus when trying to argue that fewer colors will not suffice, we look for odd cycles (which require 3 colors), odd wheels (which require 4 colors), and cliques (which require as many colors as the number of vertices in the clique). Below is a summary of our discussion so far regarding lower bounds for the chromatic number of a graph.

Special Classes of Graphs with known $\chi(G)$

- $\chi(C_n) = 2$ if n is even $(n \geq 2)$

- $\chi(C_n) = 3$ if n is odd $(n \geq 3)$

- $\chi(K_n) = n$

- $\chi(W_n) = 4$ if n is odd $(n \geq 3)$

One note of caution: a graph can have a chromatic number that is much larger than its clique size. In fact, Jan Mycielski showed that there exist graphs with an arbitrarily large chromatic number yet have a clique size of 2. We often refer to graphs with $\omega(G) = 2$ as *triangle-free*. Mycielski's proof provided a method for finding a triangle-free graph that requires the desired number of colors.

Example 6.2 Mycielski's Construction is a well-known procedure in graph theory that produces triangle-free graphs with increasing chromatic numbers. The idea is to begin with a triangle-free graph G where $V(G) = \{v_1, v_2, \ldots, v_n\}$ and add new vertices $U = \{u_1, u_2, \ldots, u_n\}$ so that $N(u_i) = N(v_i)$ for every i; that is, add an edge from u_i to v_j whenever v_i is adjacent to v_j. In addition, we add a new vertex w so that $N(w) = U$; that is, add an edge from w to every vertex in U. The resulting graph will remain triangle-free but need one more color than G. If you perform enough iterations of this procedure, you can obtain a graph with $\omega(G) = 2$ and $\chi(G) = k$ for any desired value of k.

Consider G to be the complete graph on two vertices, K_2, which is clearly triangle free and has chromatic number 2, as shown below.

After the first iteration of Mycielski's Construction, we get the graph shown below on the left. Notice that u_1 has an edge to v_2 since v_1 is adjacent to v_2. Similarly, u_2 has an edge to v_1. In addition, w is adjacent to both u_1 and u_2. The graph on the right below is an unraveling of the graph on the left. Thus we have obtained C_5, which we know needs 3 colors.

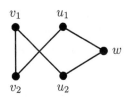 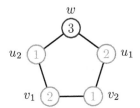

After the second iteration, we obtain the graph shown below. The outer cycle on 5 vertices represents the graph obtained above in the first iteration. The inner vertices are the new additions to the graph, with u_1 adjacent to v_2 and v_5 since v_1 is adjacent to v_2 and v_5. Similar arguments hold for the remaining u-vertices and the center vertex w is adjacent to all of the u-vertices. A coloring of the graph is shown below on the right. Note that the outer cycle needs 3 colors, as does the group of u-vertices. This forces w to use a fourth color. In addition, no matter which three vertices you choose, you cannot find a triangle among them, and so the graph remains triangle-free.

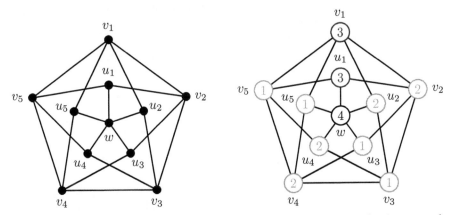

If we continue this procedure through one more step, we obtain a graph needing 5 colors with a clique size of 2.

Although Mycielski's Construction should warn you not to rely too heavily on the clique size of a graph, most real world applications have a chromatic number close to their clique size.

The discussion so far has focused on lower bounds for the chromatic number of a graph. However, when searching for an optimal coloring, it is often useful to know upper bounds as well. Combined with the lower bounds we found above, we get a nice range from which to narrow our search for an optimal coloring. Perhaps the most useful of results for upper bounds is the following theorem due to the English mathematician Rowland Leonard Brooks and published in 1941.

Theorem 6.5 (Brooks' Theorem) Let G be a connected graph and Δ denote the maximum degree among all vertices in G. Then $\chi(G) \le \Delta$ as long as G is not a complete graph or an odd cycle. If G is a complete graph or an odd cycle then $\chi(G) = \Delta + 1$.

The reasoning behind Brooks' Theorem is that if all the neighbors of x have been given different colors, then one additional color is needed for x. If x has the maximum degree over all vertices in G, then we have used $\Delta + 1$ colors for x and its neighbors. Perhaps more surprising is that unless a graph equals K_n or C_m (for an odd m), the neighbors of the vertex of maximum degree cannot all be given different colors and so the bound tightens to Δ. For example, the third graph from Mycielski's construction in Example 6.2 has a maximum degree of 5. Since this graph is neither a complete graph nor an odd cycle (although it does contain an odd cycle), we know the chromatic number is at most 5. In addition, since the clique size is 2, we know the chromatic number must be at least 2. As we showed above, the correct value was 4.

One final note of caution: although combining Brooks' Theorem with the known chromatic numbers of specific subgraph structures (such as complete graphs and odd wheels) narrows the range of possible values of the chromatic number of a graph, in practice the range between the maximum degree and the clique size of a graph can be quite large. The results we have discussed so far are simply tools to aid in further examination of a graph's structure.

6.3   Coloring Strategies

The bounds above provide starting points for determining the range in which to search for a proper k-coloring of a graph. The process for finding a minimum coloring is not trivial, and in fact belongs to a class of problems known as NP-Complete (see Section 7.1), though we will discuss some strategies for determining the chromatic number of a graph. These strategies are split into two categories: those for when the entire graph is available to you, which we will call general strategies, and those for when only pieces of the graph are visible at one time, called on-line coloring.

General Strategies

In our discussion of Brooks' Theorem, we noted that if every neighbor of a vertex has a different color, then one additional color would be needed for that vertex. This implies that large degree vertices are more likely to increase the value for the chromatic number of a graph and thus should be assigned a color earlier rather than later in the process. In addition, it is better to look for locations in which colors are forced rather than chosen; that is, once an initial vertex is given color 1, look for cliques within the graph containing that vertex as these have very clear restrictions on assigning future colors.

Example 6.3 Every year on Christmas Eve, the Petrie family compete in a friendly game of Trivial Pursuit. Unfortunately, due to longstanding disagreements and the outcome of previous years' games, some family members are not allowed on the same team. The table below lists the ten family members competing in this year's Trivial Pursuit game. An entry of N in the table indicates people who are incompatible. Model the information as a graph and find the minimum number of teams needed to keep the peace this Christmas.

	Betty	Carl	Dan	Edith	Frank	Henry	Judy	Marie	Nell	Pete
Betty	.	.	N	.	.	.	.	N	N	N
Carl	.	.	N	N	.	.	.	N	.	.
Dan	N	N	.	N	.	.	.	N	N	N
Edith	.	N	N	.	.	.	.	N	.	.
Frank	.	.	.	.	.	N	N	.	.	N
Henry	.	.	.	.	N	.	.	.	.	N
Judy	.	.	.	.	.	N	.	.	N	N
Marie	N	N	N	N	.	.	.	.	N	N
Nell	N	.	N	.	.	.	N	N	.	N
Pete	N	.	N	.	N	N	N	N	N	.

Solution: Each person will be represented by a vertex and an edge indicates two people who are incompatible, as shown below.

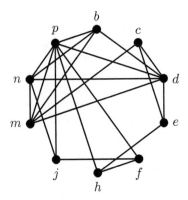

Colors will be assigned to the vertices, where each color represents a Trivial Pursuit team. At our initial step, we want to find a vertex of highest degree (p) and give it color 1. Once p has been assigned a color, we look at its neighbors with high degree as well, namely d (degree 6), m and n (both of degree 5).

These four vertices are also all adjacent to each other (forming a K_4 shown in blue below) and so must use three additional colors.

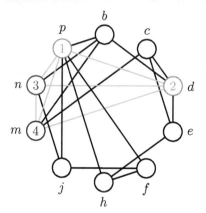

Finally, b has the next highest degree (4) and is also adjacent to all the previously colored vertices (forming a K_5) and so a fifth color is needed. The remaining vertices all have degree 3 and can be colored without introducing any new colors. One possible solution is shown below.

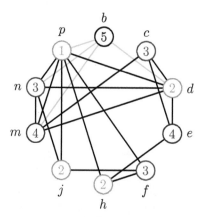

This solution translates into the following teams:

Team	Members		
1	Pete		
2	Dan	Henry	Judy
3	Carl	Frank	Nell
4	Edith	Marie	
5	Betty		

The coloring obtained in Example 6.3 was not unique. There are many ways to find a proper coloring for the graph; however, every proper coloring would need at least five colors.

In terms of the graph model (forming teams) does the solution above seem fair? Often we are not only looking for the minimal k-coloring, but also one that adds in a notion of fairness.

Definition 6.6 An *equitable coloring* is a minimal proper coloring of G so that the number of vertices of each color differs by at most one.

By this definition, the final coloring from Example 6.3 is not equitable. Note that not all graphs have equitable coloring using exactly $\chi(G)$ colors.

Example 6.4 Find an equitable coloring for the graph from Example 6.3.

Solution: We begin with the 5-coloring obtained in Example 6.3. Note that colors 2 and 3 are each used three times, color 4 twice, and colors 1 and 5 each once. This implies we should try to move one vertex each from color 2 and color 3 and assign either color 1 or color 5. One possible solution is shown below.

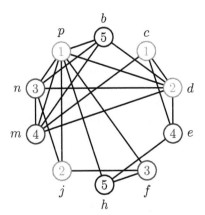

A common strategy for coloring is to begin by finding all vertices that can be given color 1, and once that is done find all the vertices that can be given color 2, and so on. The problem with this strategy is that you may choose to give vertex x color 1 which can necessitate the addition of a new color for vertex y when if x was given color 2, then y could be colored using one of the previously used colors. Instead we should always focus on locations that *force* specific colors to be used rather than *choose* which color to use. Below is a summary of the coloring strategies we have discussed so far.

General Coloring Strategies

- Begin with vertices of high degree.

- Look for locations where colors are forced (cliques, wheels, odd cycles) rather than chosen.

- When these strategies have been exhausted, color the remaining vertices while trying to avoid using any additional colors.

On-line Coloring

On-line coloring differs from a general coloring in that the vertices are examined one at a time (hence they are seen in a linear manner, or "on a line"). Often, we are restricted to situations where portions of the graph are visible at different times and so a vertex must be assigned a color without all the information available. The notion of an on-line coloring relies on a specific type of subgraph, called an *induced subgraph*.

Definition 6.7 Given a graph $G = (V, E)$, an ***induced subgraph*** is a subgraph $H = (V', E')$ where $V' \subseteq V$ and every available edge from G between the vertices in V' is included.

Another way of thinking of an induced subgraph is that we remove all edges from the graph that do not have both endpoints in the vertex set V'.

Example 6.5 Consider the graph G below. Find an induced subgraph H_1 and a subgraph H_2 that is not an induced subgraph both with the vertex set $V' = \{a, b, c, f, g, i\}$.

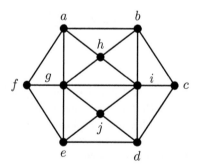

Solution: The graph H_1 on the left below is induced since every edge from G amongst the vertices in V' is included. The graph H_2 on the right below is not induced since some of the available edges are missing (namely, ab, af, ci and gi).

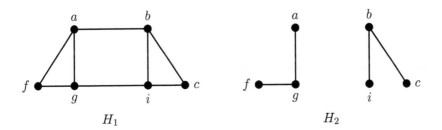

The main reason we need induced subgraphs for coloring problems is that if we took any subgraph and colored it, we may be missing edges that would indicate two vertices need different colors in the larger graph. On-line coloring algorithms require a vertex to be colored based only upon the induced subgraph containing that vertex and the previously colored vertices.

Definition 6.8 Consider a graph G with the vertices ordered as $x_1, x_2, \ldots, x_n$. An **on-line algorithm** colors the vertices one at a time where the color for x_i depends on the induced subgraph H_i which consists of the vertices up to and including x_i (so $V(H_i) = \{x_1, x_2, \ldots, x_i\}$). The maximum number of colors a specific algorithm $\mathcal{A}$ uses on any possible ordering of the vertices is denoted $\chi_{\mathcal{A}}(G)$.

Many different on-line algorithms exist, some of which can be quite complex. Mathematicians are often interested in finding an on-line algorithm that works well on a specific type of graph, or in showing how the underlying structure of specific types of graphs limits the performance of any on-line algorithm. We will focus on a greedy algorithm called *First-Fit* that uses the first available color for a new vertex.

First-Fit Coloring Algorithm

Input: Graph G with vertices ordered as $x_1, x_2, \ldots, x_n$.

Steps:

1. Assign x_1 color 1.

2. Assign x_2 color 1 if x_1 and x_2 are not adjacent; otherwise, assign x_2 color 2.

3. For all future vertices, assign x_i the least number color available to x_i in H_i; that is, give x_i the first color not used by any neighbor of x_i that has already been colored.

Output: Coloring of G.

One of the benefits of First-Fit is the ease with which it is applied. When a new vertex is encountered, we simply need to examine its neighbours that have already been colored and give the new vertex the least color available.

Example 6.6 Apply the First-Fit Algorithm to the graph from Example 6.3 if the vertices are ordered alphabetically.

Solution: To emphasize that only some of the graph is available at each step of the algorithm, only the edges to previously considered vertices will be drawn.

Step 1: Color b with 1.

Step 2: Color c with 1 since b and c are not adjacent.

Step 3: Color d with 2 since d is adjacent to a vertex of color 1.

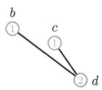

Step 4: Color e with 3 since e is adjacent to a vertex of color 1 (c) and a vertex of color 2 (d).

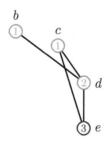

Step 5: Color f with 1 since f is not adjacent to any previous vertices.

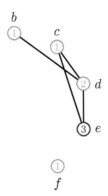

Step 6: Color h with 2 since h is adjacent to a vertex of color 1 (f).

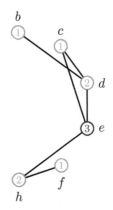

Step 7: Color j with 2 since j is adjacent to a vertex of color 1 (f).

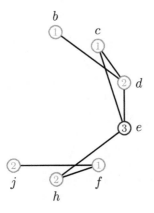

Step 8: Color m with 3 since m is adjacent to vertices of color 1 (b, c) and a vertex of color 2 (d).

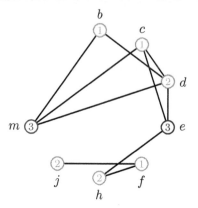

Step 9: Color n with 4 since n is adjacent to vertices of color 1, 2, and 3.

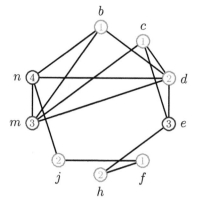

Step 10: Color p with 5 since p is adjacent to vertices of color 1, 2, 3, and 4.

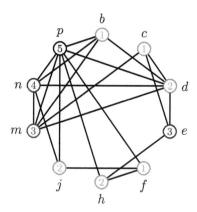

As the example above shows, First-Fit can perform quite well given the proper ordering. Unfortunately, given the right graph and wrong order of the vertices, First-Fit can perform remarkably poorly, as seen below.

Example 6.7 A vertex will be revealed one at a time, along with any edges to previously seen vertices. The First-Fit Algorithm will be applied in the order the vertices are seen.

Step 1: The first vertex is v_1. It is given color 1.

v_1

Step 2: The second vertex is v_2. Since there is no edge to v_1, it is also given color 1.

v_1

①
v_2

Step 3: The third vertex, v_3, has an edge to v_2 and so must be assigned color 2.

Step 4: The fourth vertex, v_4, has an edge to v_1 but not v_3 and so is also given color 2.

Step 5: The fifth vertex, v_5, is adjacent to a vertex of color 1 (v_2) and a vertex of color 2 (v_4). It must be assigned color 3.

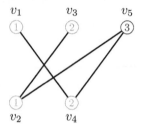

Step 6: The sixth vertex, v_6, is also adjacent to a vertex of color 1 (v_1) and a vertex of color 2 (v_3) but not a vertex of color 3 so it can also be given color 3.

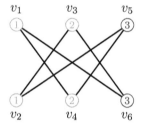

If we continue in this fashion, after $2t$ steps we will have used t colors.

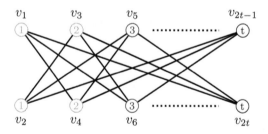

However, this graph is bipartite and every bipartite graph can be colored using 2 colors, as shown below,

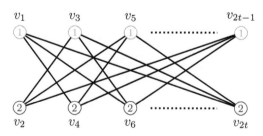

First-Fit is in fact one of the worst performers for on-line algorithms since it uses very little knowledge of how the previous vertices were treated. When evaluating a coloring algorithm's performance, we are asking what is the most number of colors the algorithm could use on any ordering of the vertices, denoted $\chi_{FF}(G)$. It is possible for an on-line algorithm (even First-Fit) to provide an optimal coloring, but in most cases this does not occur. The next section investigates a special class of graph where on-line algorithms perform remarkably well.

6.4   Perfect Graphs

Mycielski's construction demonstrated that the clique size and chromatic number of a graph can be quite far apart. Graphs where these numbers are equal, not only for the entire graph but for all induced subgraphs, comprise a special class of graphs called *perfect graphs*.

Definition 6.9 A graph G is called **perfect** if $\chi(H) = \omega(H)$ for every induced subgraph H of G.

Notice that this description did not state all subgraphs satisfy $\chi(H) = \omega(H)$, but only those that are *induced subgraphs*. There are many types of graphs that fall under the class of perfect graphs. We will investigate two of these, interval graphs and tolerance graphs, as they appear when modeling real world problems as a graph coloring problem.

Interval Graphs

Look back at the example proposed at the beginning of this chapter. Five student groups need to schedule meetings while using the fewest rooms possible. The fact that we are wanting to minimize something (number of rooms) and there are conflicts among different groups (overlapping times) should indicate this problem can be modeled using graph coloring. In fact, the colors represent the rooms (since this is what we are minimizing), the vertices represent the groups (since the groups are assigned a room), and the edges indicate when two groups have overlapping times (forcing the use of two different rooms). Graphs where the vertices represent intervals of time are called *interval graphs*.

Definition 6.10 A graph G is an **interval graph** if every vertex can be represented as a finite interval and two vertices are adjacent whenever the corresponding intervals overlap; that is, for every vertex x there exists an interval I_x and xy is an edge in G if $I_x \cap I_y \neq \emptyset$.

Beyond their applicability to real world problems, interval graphs are nice graphs for coloring problems as they fall under the class of perfect graphs. Thus to demonstrate that you have found the chromatic number and a better coloring does not exist, it is enough to find the clique size of the graph.

Example 6.8 Five student groups are meeting on Saturday, with varying time requirements. The staff at the Campus Center need to determine how to place the groups into rooms while using the fewest rooms possible. The times required for these groups is shown in the table below. Model this as a graph and determine the minimum number of rooms needed.

Student Group	Meeting Time
Agora	13:00 – 15:30
Counterpoint	14:00 – 16:30
Spectrum	9:30 – 14:30
Tupelos	11:00 – 12:00
Upstage	11:15 – 15:00

Solution: First we display the information in terms of the intervals. Although this step is not necessary, sometimes the visual aids in determining which vertices are adjacent.

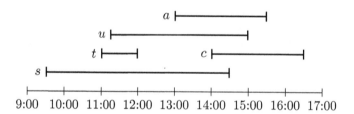

Below is the graph where each vertex represents a student group and two vertices are adjacent if their corresponding intervals overlap.

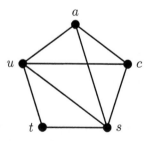

A proper coloring of this graph is shown below. Note that four colors are required since there is a K_4 subgraph with a, c, s and u.

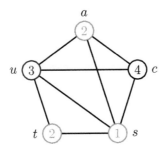

It should be noted that in most applications of interval graphs, you are given the intervals and must form the graph. A much harder problem is determining if an interval representation of a graph exists and then finding one.

On-line coloring algorithms perform nicely on interval graphs. In particular, if the interval representation is known and the vertices are ordered by starting time of their intervals, then First-Fit will produce a coloring using exactly $\chi(G)$. However, if the intervals are not ordered by their starting time or a different on-line algorithm is used, the optimal coloring may not be found.

Example 6.9 Ten customers are buying tickets for various trips along the Pacific Northwest train route shown on the right. Each person must be assigned a seat when a ticket is purchased and you only know which seats have been previously assigned. Using a random assignment of seat numbers as the information becomes available, find a way to minimize the number of seats required.

Solution: Each step indicates when a new person buys a ticket. Their representative vertex must be assigned a color before moving to the next step.

Step 1: Cathy buys her ticket for Bellingham to Edmonds. She is assigned seat 1, as shown in the graph below.

① c

Step 2: Fiona buys her ticket next for Bellingham to Renton. Since she and Cathy will be on the train at the same time, she must have a different seat. She is assigned seat 2.

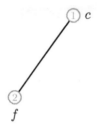

Step 3: Next, Greg buys a ticket for Vancouver to Mount Vernon. His trip overlaps with those of both Cathy and Fiona, necessitating the use of another seat. He is assigned seat 3.

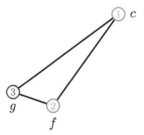

Step 4: Ben buys the next ticket for Portland to Oregon City. Since his trip does not overlap with any of the earlier purchased tickets, he can be assigned any seat. We choose seat 1.

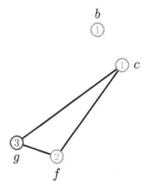

Step 5: Next Ingrid buys a ticket for Oregon City to Albany. Similar to Ben, she can also be assigned any seat. We choose seat 3.

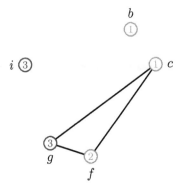

Step 6: Jessica buys the next ticket for Albany to Eugene. As in the previous two steps, she can be assigned any seat. We choose seat 2.

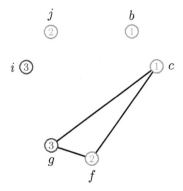

Step 7: Howard buys a ticket for Centralia to Eugene. Since his route overlaps with those of Ben, Jessica and Ingrid, he must be assigned to a new seat, number 4.

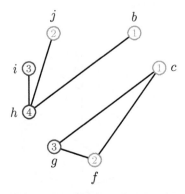

Step 8: Dana buys the next ticket for Renton to Tacoma. He can be assigned any seat, so we choose seat 3.

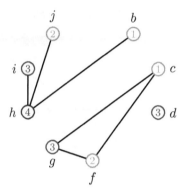

Step 9: Aiden buys a ticket for Seattle to Oregon City. Since his trip overlaps with someone in seats 1 through 4, we need a new seat for him, number 5.

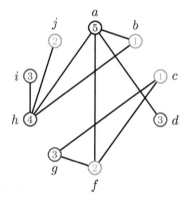

Step 10: Emily is the last person to buy a ticket for a trip from Everett to Kelso. Since her trip overlaps with someone in each of the previously assigned seats, we need a sixth seat for Emily.

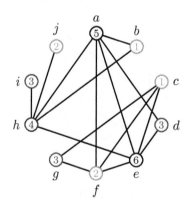

In the example above, we did not use First-Fit but rather a more random choice for an on-line algorithm. In Exercise 6.6 First-Fit is applied to this graph using the same order of the vertices, resulting in an improvement on the number of colors (or seats). However, neither of these algorithms produces an optimal coloring as the chromatic number for this graph is 3 since the clique size is 3 (for example, there is a K_3 with vertices a, b and h), and since this is an interval graph we know $\chi(G) = \omega(G)$. One optimal coloring is shown below.

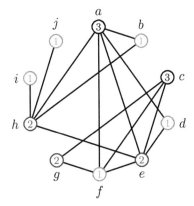

In general, First-Fit can be shown to be no worse than roughly $5\chi(G)$ on interval graphs (which means First-Fit will never use more than five times the optimal) and there exists an on-line algorithm that uses at most $3\chi(G)$ colors on interval graphs (see [28],[29]). Though this may seem quite large, recall that in Example 6.7 we had a graph with $2t$ vertices using t colors when only 2 were needed, implying a performance of $\frac{n}{4}\chi(G)$ for bipartite graphs with n vertices.

Tolerance Graphs

Consider the scenario from Example 6.8 where five student groups needed to schedule meetings. Due to the overlapping times, we needed four rooms to accommodate the groups' time requirements. But what if there were not four rooms available? Most groups would prefer to shorten their meeting by a small amount of time if it meant they could still hold their meeting. This idea of leeway can be modeled using a *tolerance graph*.

Definition 6.11 A graph G is a **tolerance graph** if every vertex can be represented as a finite interval with a non-negative tolerance so that two vertices are adjacent whenever the corresponding intervals overlap by at least the smaller of their two tolerances; that is, for every vertex x there exists an interval I_x with tolerance t_x and xy is an edge if $I_x \cap I_y \neq \emptyset$ and $|I_x \cap I_y| \geq \min\{t_x, t_y\}$.

Tolerance graphs fall under the class of perfect graphs, and as such determining the chromatic number boils down to finding the clique size of the graph. As with interval graphs, determining if a graph is a tolerance graph can be quite complex but most applications have a tolerance representation already provided.

Example 6.10 Each of the groups from Example 6.8 has been asked to submit some leeway (in minutes) for their meeting so that the staff at the Student Center can schedule more activities. These are listed below along with the desired meeting times. Determine the minimum number of rooms needed for all five groups to hold their meetings when tolerances are taken into account.

Student Group	Meeting Time	Leeway
Agora	13:00 – 15:30	30
Counterpoint	14:00 – 16:30	40
Spectrum	9:30 – 14:30	45
Tupelos	11:00 – 12:00	5
Upstage	11:15 – 15:00	15

Solution: As in Example 6.8, we begin by drawing the intervals representing each group. Note that the gray bars indicate the tolerance (or leeway) for the meeting time of each group.

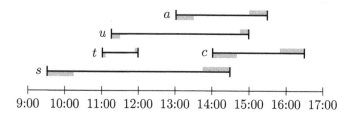

To form the graph, we only draw an edge if the two intervals overlap by more than the smallest of the two tolerances. From the graphic above, this means that the intervals overlap beyond the gray bars that indicate the tolerances, and we can produce the graph below. Note that the graph is almost identical to the graph from Example 6.8, only now s and c are not adjacent since the size of the intersection of their intervals is 30 minutes, whereas the tolerances are 45 and 40 minutes, respectively.

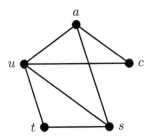

The graph now only needs 3 colors, as shown below. We know fewer colors will not suffice since there is a K_3 subgraph (for example between a, c, u). This implies the five groups can be placed into 3 rooms.

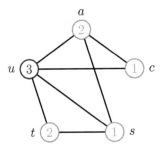

Like interval graphs, on-line algorithms can be shown to perform rather well on tolerance graphs, though the results require more information as to how the tolerances relate to the length of the interval. For further information, see [18] or [26].

Example 6.11 Draw a graph representing the intervals below with the tolerances shown in gray. Determine the chromatic number.

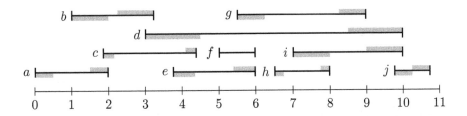

Solution: Below is the graph where every vertex represents an interval. Notice that even though intervals b and d overlap, the size of their intersection is less than the minimum tolerance and so there is no edge between these vertices in the graph. Also, the lack of gray bars on f indicates that the tolerance for f is 0, and so vertex f will be adjacent to any vertex whose interval overlaps that of f.

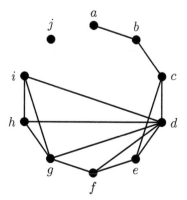

A minimum coloring is given below. We know the chromatic number is 4 since there is a K_4 subgraph (among vertices d, g, h and i).

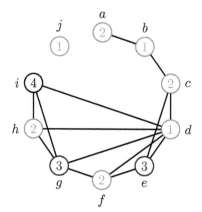

6.5   Weighted Coloring

Consider the following scenario:

> Ten families need to buy train tickets for an upcoming trip. The families vary in size but each of them needs to sit together on the train.

Determine the minimum number of seats needed to accommodate the ten family trips.

This problem should sound very similar to Example 6.9 where colors were representing seats and the vertices were intervals of time indicative of when a person was on the train. Here, we are still interested in a graph coloring, but now each vertex represents a family and so has a size associated with it. In previous chapters, we used weights on the edges of a graph to indicate distance, time, or cost. For graph coloring models, weighted edges would have very little meaning. Instead, we will assigning a weight to each vertex, and finding a proper coloring will be referred to as a *weighted coloring*.

Definition 6.12 Given a weighted graph $G = (V, E, w)$, where w assigns each vertex a positive integer, a proper **weighted coloring** of G assigns each vertex a set of colors so that

(i) the set consists of consecutive colors (or numbers);

(ii) the number of colors assigned to a vertex equals its weight; and

(iii) if two vertices are adjacent, then their set of colors must be disjoint.

Note that in some publications, weighted colorings are referred to as interval colorings (since an interval of colors is being assigned to each vertex). To avoid the confusion between interval colorings and interval graphs, we use the term weighted coloring.

Before we tackle the train example, we will look at a smaller graph with a weighted coloring. We will, for the most part, use the same strategies for finding a minimum weighted coloring as we did above for unweighted coloring. The biggest change will be to focus on locations that have large weighted cliques, which is found by adding the weights of the vertices within a complete subgraph. Thus if we have two different cliques on three vertices, one with total weight 8 and the other with total weight 10, we should initially focus on the one with the higher total weight.

Example 6.12 Find an optimal weighted coloring for the graph below where the vertices have weights as shown below.

$w(a) = 2$
$w(b) = 1$
$w(c) = 4$
$w(d) = 2$
$w(e) = 2$
$w(f) = 4$

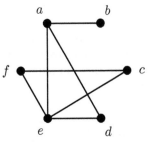

Solution: Note that a, d and e form a K_3 with total weight 8, and c, e and f

form a K_3 with total weight 10. We begin by assigning weights to the vertices from the second K_3, and since e appears twice we assign it the first set of colors $\{1, 2\}$. From there we are forced to use another 4 colors on f ($\{3, 4, 5, 6\}$) and another 4 colors on c ($\{7, 8, 9, 10\}$).

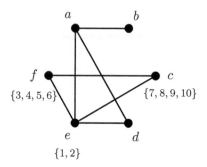

We can now fill in the remaining three vertices using the colors 1 through 10.

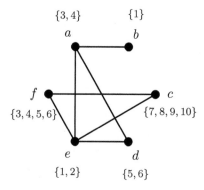

As the example above demonstrates, we focus less on a vertex degree and more on the vertex weight when we are searching for a minimal weighted coloring. This is in part due to the need for the set of colors to be consecutive. If a vertex has high weight, then it needs a larger range from which to pick the set of colors, whereas a vertex with large degree but a small weight may be able to squeeze in between the sets of colors of its neighbors.

Example 6.13 Suppose the ten families needing train tickets have the same underlying graph as that from Example 6.9 and the size of each family is noted below. Determine the minimum number of seats needed to accommodate everyone's travels.

$w(a) = 2$
$w(b) = 5$
$w(c) = 2$
$w(d) = 1$
$w(e) = 4$
$w(f) = 3$
$w(g) = 1$
$w(h) = 3$
$w(i) = 4$
$w(j) = 5$

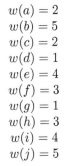
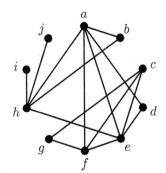

Solution: As with the example above, we begin by looking for the largest total weight for a clique. Since the clique size is three, we want to find K_3 subgraphs with high total weight. The largest of these is with a, b, and h with total 10. Since b has the largest weight among these, we give it colors 1 through 5, assign a colors 6 and 7, and colors 8, 9, and 10 to h.

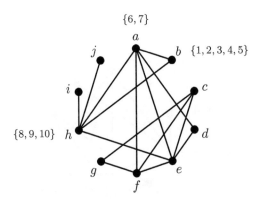

The next two cliques with a high total weight (of 9) are formed by a, e, f and c, e, f. Since a has already been assigned colors, we begin with the first clique. We can fill in the colors for e and f without introducing new colors, as shown below. Note that since e is adjacent to both a and h it could only use four consecutive colors chosen from those used on b.

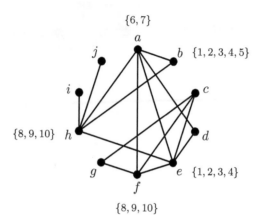

At this point we can fill in the colors for c by choosing two consecutive colors from those available (5, 6, and 7). We also fill in the colors for i and j by choosing the correct number of consecutive colors from any of 1 to 7.

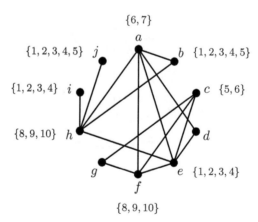

Finally, we need one color each for d and g. These can be any color not already used by one of their neighbors.

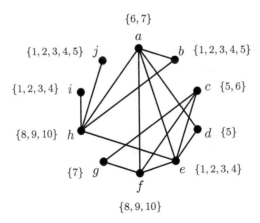

On-line algorithms, and specifically First-Fit, can also be used for weighted colorings. It should not come to much surprise that on-line algorithms generally use more colors than when the entire graph can be seen. In Exercise 6.10 you are asked to use First-Fit on the example above.

The combination of interval graphs, on-line algorithms, and weighted colorings have been extensively studied due to a very specific application, called Dynamic Storage Allocation, or DSA. The storage allocation refers to assigning variables to locations within a computer's memory, where each variable has a size associated to it. This can be thought of as the weight of a vertex. The location a variable is assigned is the set of colors a vertex is given. The dynamic part of DSA refers to variables being in use for specific intervals of time and only the previously used (or in-use) variables locations are know. Thus each vertex is represented by an interval and the coloring must use an on-line algorithm. In total, DSA can be modeled as an on-line coloring of a weighted interval graph. Modifications to the known performance of algorithms for interval graphs can be used to provide limitations on DSA performance. In addition, DSA has been generalized to account for leeway of storage, making use of tolerance graphs in place of interval graphs. For further information, see [25], [26], and [27].

6.6 Exercises

6.1 Find a coloring of the map of the United States. Explain why four colors are necessary.

6.2 Find the chromatic number for each of the graphs below. Include an argument why fewer colors will not suffice.

(a)

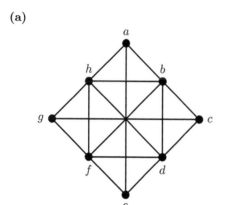

(b)

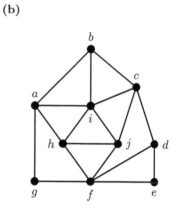

(c)

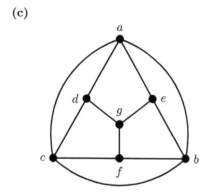

(d)

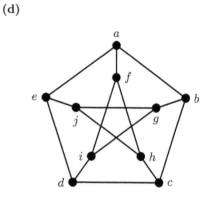

(e)

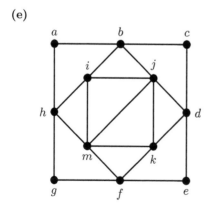

(f)

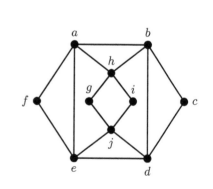

6.3 (From [39]) A set of solar experiments is to be made at various observatories and each experiment is to be repeated for several years, as shown in the table below. Each experiment begins on a given day of the year and ends on a different given day and an observatory can perform only one experiment at a time. Determine the minimum number of observatories required to perform a given set of experiments annually.

Experiment	Start Date	End Date
A	Sept 2	Jan 3
B	Oct 15	Mar 10
C	Nov 20	Feb 17
D	Jan 23	May 30
E	Apr 4	July 28
F	Apr 30	July 28
G	June 24	Sept 30

6.4 Ten students in the coming semester will be taking the courses shown in the table below.

Course	Students				
Physics	Arnold	Ingrid	Fred	Bill	Jack
Mathematics	Eleanor	Arnold	Herb		
English	Arnold				
Geology	Carol	Bill	Fred	Herb	
Business	George	Eleanor	Carol		
Statistics	David	Ingrid	George		
Economics	Ingrid	Jack			

(a) Draw a graph modeling these conflicts.
(b) How many time periods must be allowed for these students to take the courses they want without conflicts? Include an argument why fewer time periods will not suffice.
(c) The following semester, all the students except David plan to take second courses in the same subject. David decides not to take further courses in Statistics. How many time periods will then be required?

6.5 The seven committees from Exercise 5.8 need to schedule their weekly meeting. Based on the membership lists (shown below), determine the number of time slots needed so each person can make his or her committee meeting.

Committee	Members			
Benefits	Agatha	Dinah	Evan	Vlad
Computing	Evan	Nancy	Leah	Omar
Purchasing	George	Vlad	Leah	
Recruitment	Dinah	Omar	Agatha	
Refreshments	Nancy	George		
Social Media	Evan	Leah	Vlad	Omar
Travel Expenses	Agatha	Vlad	George	

6.6 Using the same order of the vertices, apply First-Fit to the graph from Example 6.9.

6.7 Draw a graph representing the same intervals from Example 6.11 where all the tolerances are 0 (and so we are back to an interval graph). Find the chromatic number of this graph.

6.8 Below is a collection of meetings that need to be assigned to conference rooms. Each group has identified when it would like to meet and how long it needs. In addition, each group has given the organizers a little leeway in the amount of time it is willing to cut short its meeting if the room is needed for another group. Model this information as a graph and determine how many conference rooms are needed.

Organization	Time	Leeway
Adam's Apples	8:30 – 9:30	10 minutes
Brain Teasers	9:00 – 11:00	45 minutes
Cookie Club	9:45 – 12:00	30 minutes
Disaster Readiness	10:00 – 1:00	45 minutes
Edison Enthusiasts	11:15 – 12:45	5 minutes
Fire Chiefs	12:30 – 3:30	30 minutes
Gary's Golfers	2:00 – 3:15	5 minutes
Helix Doubles	2:15 – 4:00	1 hour

6.9 Find an optimal weighted coloring for each of the graphs from Exercise 6.2 with the weights as shown below.

Vertex	Weight in graph 6.2					
	(a)	(b)	(c)	(d)	(e)	(f)
a	2	4	1	3	3	3
b	3	2	4	3	3	5
c	2	5	3	2	4	3
d	2	3	2	2	3	2
e	4	2	3	4	5	4
f	2	3	4	3	3	2
g	3	1	2	2	4	2
h	1	3	·	3	4	3
i	·	2	·	1	1	3
j	·	1	·	1	4	3
k	·	·	·	·	3	·
m	·	·	·	·	5	·

6.10 Using the same order of vertices as from Example 6.9, apply First-Fit to find a weighted coloring of the graph from Example 6.13.

Projects

6.11 This chapter focused on vertex coloring, with a brief discussion of weighted colorings. Another modification exists where instead of weights assigned to vertices, each vertex is given a specific list of colors from which to choose (see Section 7.7 for coloring the edges of a graph). More formally, each vertex x is given a list $L(x)$ of colors and a proper *list coloring* assigns to x a color from its list $L(x)$ so that no two adjacent vertices are given the same color. Given any possible collection of lists, each of size k, to the vertices of a graph G, if a proper list coloring exists then G is k-*choosable*. The minimum value for k for which G is k-choosable is called the *choosability* of G and denoted $ch(G)$.

(a) Find a list coloring for select graphs from Exercise 6.2 with the lists as shown on the next page.

(b) Can the graph 6.2(a) be list colored if each vertex is given the same list of size 4? size 3? Explain your answer.

(c) Compare your answer from (b) with your coloring in (a). Explain why the answers for lists of size 3 differ.

(d) Explain why $ch(G) \geq \chi(G)$ for all graphs G.

(e) Explain why $ch(G) \leq \Delta(G) + 1$ for all graphs G.

	Lists for graph 6.2		
Vertex	**(a)**	**(c)**	**(e)**
a	$\{1,2,3\}$	$\{1,2,3\}$	$\{1,2\}$
b	$\{4,5,6\}$	$\{2,3,4\}$	$\{2,3\}$
c	$\{1,2\}$	$\{3,4\}$	$\{3,4\}$
d	$\{2,4\}$	$\{1,4\}$	$\{1,2\}$
e	$\{4,7,8\}$	$\{1,3,4\}$	$\{1,3,4\}$
f	$\{5,8\}$	$\{2,3,4\}$	$\{1,2,4\}$
g	$\{3\}$	$\{1,4\}$	$\{1,2\}$
h	$\{2,4,6\}$	$\cdot$	$\{2,3\}$
i	$\cdot$	$\cdot$	$\{4,5\}$
j	$\cdot$	$\cdot$	$\{1,2,3\}$
k	$\cdot$	$\cdot$	$\{3,4\}$
m	$\cdot$	$\cdot$	$\{2,4\}$

Chapter 7

Additional Topics

Graph Theory is a rich field of mathematics. Although it dates back to Euler's paper in 1736, much of the work has been done in the last century. The first six chapters of this book focus on the more algorithmic and applied areas of Graph Theory. The purpose of this chapter is to expand on some concepts that appeared in previous chapters and explore additional topics that do not warrant an entire chapter.

The sections to follow are independent of each other, though they rely on terms and topics defined earlier in the book. References to previous sections or topics appear as needed. Each section concludes with limited exercises for further practice and deepening your understanding.

7.1 Algorithm Complexity

In Chapter 2 we spent time discussing efficiency of algorithms, in particular the challenges of using Brute Force to solve a Traveling Salesman problem. At other times throughout this book, algorithm efficiency and performance were mentioned to explain why a specific problem was difficult to solve (see 3.1, 4.2, 4.3, and 6.3). This section will elaborate on what algorithm efficiency means and how mathematicians determine what makes one algorithm more efficient than another.

The algorithms we have studied throughout this book have been written so that *you*, the reader, could perform the necessary computations to find a solution. However, all of these algorithms can be written in such a way as to have a computer perform the calculations and provide the answer. When doing so, the information from a graph needs to be encoded in such a way as to allow the algorithm to pull the requisite information needed for the computations. For example, the Traveling Salesman Problem would need as an input the name of the n cities and the weights of the $\frac{n(n-1)}{2}$ edges in the complete graph K_n. It is possible for a given problem to have more than one way to encode the required information, so when evaluating the performance of an algorithm we will assume a standard method of encoding the input has been set and that this is done as efficiently as possible.

In addition, algorithm performance is generally evaluated based on a worst case scenario. Although an algorithm may quickly solve one instance of a problem, it is still possible to be quite slow for solving another instance. We will be concerned with how poorly an algorithm can perform over all possible instances and this evaluation is often given in terms of the running time. While the running time is dependent on the encoding scheme and computing power, changing either of these parameters does not change the complexity of the algorithm; that is, as we will see later in this section, improvements in computing power will not drastically affect the overall running time for an inefficient algorithm with large inputs.

When discussing the Brute Force Algorithm for the Traveling Salesman Problem in Chapter 2, a rough estimate for the number of required calculations was determined to be $\frac{(n+1)!}{2n}$. The estimate is given in terms of n, which is the number of cities represented in the complete graph K_n. Essentially, algorithm performance is a function $f(n)$ whose input is the size of the graph and whose output is the number of required calculations.

Definition 7.1.1 The **performance function** $f(n)$ for a graph theory algorithm is a function whose input is n, the number of vertices in the graph, and whose output is the number of required calculations to complete the algorithm.

Using this language, the performance function for Brute Force is $f(n) = \frac{(n+1)!}{2n}$.

Consider for a moment two algorithms, one whose performance function is $n^2 + 3n + 1$ and another whose performance function is n^2. When n is small, these would give slightly different outputs for running time, but as n gets larger the addition of $3n + 1$ matters much less than the common n^2 component. In essence, we consider these two algorithms to have the same complexity due to the common highest power of their performance functions. More formally, these functions would be of the same *order*.

Definition 7.1.2 Given a function f, it has **order at most** g, denoted $f(n)$ is $O(g(n))$, if there exists a constant c and a nonnegative integer a so that $|f(n)| \leq c|g(n)|$ for all $x \geq a$.

Using this notation $f(n) = n^2 + 3n + 1$ is $O(n^2)$ since for $n \geq 1$ we have $n^2 + 3n + 1 \leq n^2 + 3n^2 + n^2 = 5n^2$ (so our constants from the definition above are $c = 5$ and $a = 1$).

Algorithms with performance functions of order at most n^m, for some integer m, are called **polynomial-time algorithms**. These are, in some sense, considered "good" algorithms as they run fairly fast even as the size of the input grows. The table below gives an analysis of how various values of m impact the running time as the size of the input grows (values are rounded slightly). Note that these were calculated in the same manner as those for the Brute Force Algorithm on page 46, where we are using the best known supercomputer at the time of publication.

n	n^2	n^5	n^{10}
10	3×10^{-15} seconds	3×10^{-12} seconds	3×10^{-8} seconds
50	7.5×10^{-14} seconds	9.5×10^{-9} seconds	0.3 seconds
100	3×10^{-13} seconds	3×10^{-7} seconds	50.5 minutes
500	7.5×10^{-12} seconds	9.5×10^{-4} seconds	938 years
1000	3×10^{-11} seconds	0.03 seconds	$960,903$ years

Notice that when the polynomial has degree 5, an algorithm with input size 1000 can still be completed in less than a second! Also, even when the polynomial has degree 10, run-time only becomes infeasible after 100 inputs. In fact, only after the input size is 250 does the run-time take longer than a year and at 200 the run-time is around 1 month.

If polynomial-time algorithms are considered "good" algorithms, what constitutes a "bad" algorithm? One such type is called an **exponential-time algorithm**, where the performance functions are of order at most k^n for some constant k. For example $f(n) = 2^n + 5n + 1$ is of order at most 2^n since for all $n \geq 0$ we have $2^n + 5n + 1 \leq 2^n + 5 \cdot 2^n + 2^n = 7 \cdot 2^n$ (so our constants from the definition above are $c = 7$ and $a = 0$). An algorithm with this performance function is considered an inefficient algorithm since small increases in the input size result in large increases in the running time, as shown in the table below.

The calculations we computed for Brute Force, gave a performance function in terms of a factorial, which is another type of inefficient algorithm called a **factorial-time algorithm** since it is of order at most $n!$. Factorial-time algorithms are in fact one of the worst in terms of their performance function. Note that there exists a more complex algorithm for solving a Traveling Salesman Problem (the Held-Karp algorithm) that has complexity $n^2 2^n$, but this still puts it in the category of inefficient algorithms since it is within the class of exponential-time algorithms.

n	2^n	3^n	$n!$
10	3×10^{-14} seconds	1.8×10^{-12} seconds	1×10^{-10} seconds
50	0.3 seconds	252 days	2.9×10^{40} years
100	1.2 million years	4.9×10^{23} years	8.9×10^{133} years
500	3×10^{126} years	3.4×10^{214} years	1.1×10^{1110} years
1000	1×10^{277} years	1.2×10^{453} years	3.8×10^{2543} years

Notice how smaller inputs for these types of algorithms quickly produce infeasible run-times. For example, even with input size 50 all of the polynomial-time algorithms in the table above had run times under a second; however, only the 2^n algorithm is reasonable. To understand the scale of the entries above, scientists believe the age of the earth to be 4.54 billion years, which is 4.54×10^9 years. Examining these algorithms with an input size of 100 shows the widening gap between polynomial-time and exponential-time or factorial-time algorithms.

All of the calculations above are based on a fixed computing power. One might ask if major improvements in computers would drastically change the feasibility of these algorithms. For example, if we are able to increase the number of calculations per second by a factor of 1000, how much faster would a n^2 algorithm run versus a 2^n algorithm? A few categories of algorithms are shown below with various factors of increase. These are all evaluated at $n = 100$.

factor increase	n^{10}	2^n	$n!$
1	50.5 minutes	1.2 million years	8.9×10^{133} years
100	30.3 seconds	$12,180$ years	8.9×10^{131} years
1000	3.03 seconds	$1,218$ years	8.9×10^{130} years
1000000	0.003 seconds	1.2 years	8.9×10^{127} years

As you can see, polynomial-time algorithms gain more of a savings from these increase factors. Is it surprising that the exponential-time and factorial-time algorithms do not gain as much of a savings? In fact, even with a computer that is a million times faster than the current best supercomputer, a factorial-time algorithm remains infeasible with only an input size of 100.

The algorithm complexities described above were chosen due to their prevalence in graph theory problems (for example, Dijkstra's Algorithm, from Section 3.1 is of order n^2). However, other time complexities exist and can be similarly analyzed. A *linear-time algorithm* has order n and a *logarithmic-time algorithm* has order $\log n$. Algorithms of these orders are even more efficient than polynomial-time algorithms and fall into the class of "good" algorithms. Additionally, some graph algorithm complexities are given in terms of both the number of vertices and the number of edges. For example, Kruskal's Algorithm is of order $m \log n$, where m is the number of edges and n is the number of vertices. However, since the number of edges in a simple graph is no more than the number of edges in a complete graph $\left(\frac{n(n-1)}{2}\right)$ any performance function using m, the number of edges, as an input can replace m with n^2 to give a rough estimate only in terms of the number of vertices.

When mathematicians and computer scientists describe algorithm complexity, they are often concerned with the overall class to which the algorithm belongs. Complexity classes are used to describe which algorithms are "good" and which are "bad." The two most commonly referenced classes are P and NP.

Definition 7.1.3 Problems that can be solved by a deterministic sequential machine using at worst a polynomial-time algorithm belong to *class* **P**. Problems for which there is no known polynomial-time solution algorithm but for which a proposed solution can be verified in polynomial-time belong to *class* **NP**.

The definition above uses the term *deterministic sequential machine* which, roughly speaking, is the equivalent of the modern computer; given a set of steps, such as those in an algorithm, a deterministic sequential machine can solve a problem and produce an output. Polynomial-time, linear-time, and logarithmic-time algorithms fall into class P, whereas exponential-time and factorial-time algorithms fall into class NP. For a more technical discussion of complexity theory, see [33].

Many of the problems discussed in this book are from class P, for example finding an Eulerian circuit (Fleury's Algorithm), a minimum spanning tree (Kruskal's or Prim's Algorithm), a shortest path (Dijkstra's Algorithm), or a maximum matching (Augmenting Path Algorithm). However, other problems are known to be in NP, such as the Traveling Salesman Problem and finding an optimal coloring. In fact, the Traveling Salesman Problem is a classic example of an NP problem that belongs to a subclass of NP, called **NP-Complete**. NP-Complete problems are considered to have equivalent complexities since if any one of the problems can be solved in polynomial-time, then all others can be solved in polynomial-time. In addition, the Steiner Tree Problem from Section 4.3 is classified as NP-Hard, a subclass of NP problems that are considered at least as hard as any NP problem.

A major question in complexity theory is whether $P = NP$ or $P \neq NP$. Another way of phrasing this is if any problem that can be verified in polynomial-time can also be solved in polynomial-time. In fact, this problem is of such importance to the fields of mathematics and computer science that it was named one of the Millennium Problems by the Clay Mathematics Institute in 2000. CMI chose seven problems believed to be of great importance to mathematics in the new millennium, and where a prize of one million dollars would be awarded for a published and verifiable solution to any of these problems. The details behind P vs. NP get quite technical, especially the deeper into the research you dive, but a final note is warranted. If $P = NP$, then our current method of Internet encryption (which is based on integer factorization, a known NP problem) would be essentially useless.

Exercises

7.1.1 Find the constants a and c, as in Definition 7.1.2, to show each of the functions below are of the order indicated.

   (a) $f(n) = n^3 + 4n + 5$ is $O(n^3)$

   (b) $f(n) = 3n^5 + n$ is $O(n^5)$

   (c) $f(n) = 2^n + n$ is $O(2^n)$

   (d) $f(n) = n! + n + 1$ is $O(n!)$

   (e) $f(n) = n! + 2^n$ is $O(n!)$

   (f) $f(n) = n^2 2^n$ is $O(2^{2n})$

7.1.2 In reference to Chapter 2, explain why Brute Force is in class NP and Nearest-Neighbor is in class P.

7.1.3 Explain algorithm complexity and the P vs. NP question in your own words.

7.2 Graph Isomorphism

Various times throughout the book, we have discussed multiple ways for drawing the same graph. In Example 1.2 we showed two different modes for drawing the graph in Example 1.1. At the time, we focused on the fact that we were dealing with the same set of vertices and verified the edge set was maintained in the new drawings. However, two graphs with distinct vertex sets can still produce the same edge relationships; more technically these graphs are called *isomorphic*.

Definition 7.2.1 Two graphs G_1 and G_2 are ***isomorphic*** if every vertex from G_1 can be paired with a unique vertex from G_2 so that corresponding edges from G_1 are maintained in G_2.

Throughout this section we will only consider simple graphs (those without multi-edges or loops). Similar definitions and results exist for multi-graphs and digraphs. A more robust definition of isomorphic uses a special function, called a bijection, between the vertices of G_1 and G_2; further discussion of bijections and isomorphisms can be found in [40].

Later we will list some of the common properties that must be maintained with isomorphic graphs. But to begin, it should be easy to name a few things that are easy to check:

- number of vertices

- number of edges

- vertex degrees

By no means is this list comprehensive, but it allows for a quick check before working on more complex ideas.

Example 7.2.1 Determine if the following pair of graphs are isomorphic. If so, give the vertex pairings; if not, explain what property is different among the graphs.

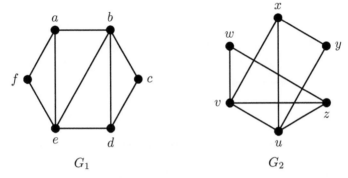

$$G_1 \qquad\qquad G_2$$

Solution: First note that both graphs have six vertices and nine edges, with two vertices each of degrees 4, 3, and 2. Since corresponding vertices must have the same degree, we know b must map to either u or v. We start by trying to map b to v. By looking at vertex adjacencies and degree, we must have e map to u, c map to w and a map to x. This leaves f and d, which must be mapped to y and z, respectively. The two charts below show the vertex pairings and checks for corresponding edges.

$V(G_1) \longleftrightarrow V(G_2)$	Edges
$a \longleftrightarrow x$	$ab \longleftrightarrow xv$ ✓
$b \longleftrightarrow v$	$ae \longleftrightarrow xu$ ✓
$c \longleftrightarrow w$	$af \longleftrightarrow xy$ ✓
$d \longleftrightarrow z$	$bc \longleftrightarrow vw$ ✓
$e \longleftrightarrow u$	$bd \longleftrightarrow vz$ ✓
$f \longleftrightarrow y$	$be \longleftrightarrow vu$ ✓
	$cd \longleftrightarrow wz$ ✓
	$de \longleftrightarrow zu$ ✓
	$ef \longleftrightarrow uy$ ✓

Since all edge relationships are maintained, we know G_1 and G_2 are isomorphic.

Example 7.2.2 Determine if the following pair of graphs are isomorphic. If so, give the vertex pairings; if not, explain what property is different among the graphs.

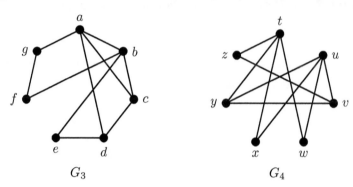

G_3 G_4

Solution: First note that both graphs have seven vertices and ten edges, with two vertices each of degrees 4 and 3, and three vertices of degree 2. As in the previous example, we know corresponding vertices must have the same degree, and so the vertices of degree 4 in G_3, a and b, must map to the vertices of degree 4 in G_4, namely t and u. However, in G_3 the degree 4 vertices (a and b) are adjacent, whereas in G_4 there is no edge between the degree 4 vertices (t and u). Thus G_3 and G_4 are not isomorphic.

The previous example illustrates that no one property guarantees two graphs are isomorphic. In fact, simply having the same number of vertices of each degree is not enough. The theorem below lists the more useful properties of graph isomorphism.

Theorem 7.2.2 Assume G_1 and G_2 are isomorphic graphs. Then G_1 and G_2 must satisfy any of the properties listed below; that is, if G_1

- is connected

- has n vertices

- has m edges

- has m vertices of degree k

- has a cycle of length k

- has an Eulerian circuit

- has a Hamiltonian cycle

then so too must G_2 (where n, m, and k are non-negative integers).

Throughout the majority of this text, we have glossed over graph isomorphism. This is in part because when modeling a problem, we only care about whether the graph in question adequately models the given information. We have not made much, if any distinction between multiple ways to draw the same set of information. In general, graph isomorphism is less applicable to real world scenarios. However, isomorphism will be briefly mentioned in Sections 7.3 and 7.6. Additionally, with respect to Section 7.1, graph isomorphism belongs in the complexity class NP, although specific graph types are known to be in class P, such as trees, interval graphs, or planar graphs (see Section 7.6).

Exercises

For each of the problems below, determine if the given pair of graphs are isomorphic. For those that are isomorphic, explicitly give the vertex correspondence and check that edge relationships are maintained. Otherwise, provide reasoning for why the pair of graphs are not isomorphic.

7.2.1

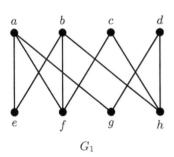

G_1

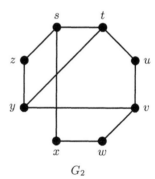

G_2

7.2.2

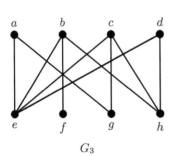

G_3

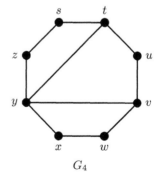

G_4

7.2.3

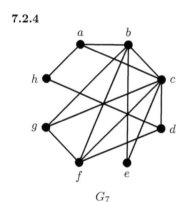

7.2.4

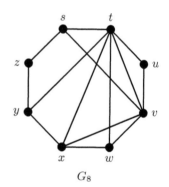

7.3   Tournaments

In Section 2.3, the concept of a digraph was introduced, where edges were replaced with directed edges, called arcs. Throughout the book, we saw how digraphs can be used to model asymmetric relationships, such as those from differing costs for travel from A to B versus B to A (Section 2.3), modeling one way streets or directional travel (Section 3.1), and precedence relationships between tasks to be scheduled (Section 3.2). This section will explore a different use of digraphs that have a very specific underlying structure, one in which a direction is added to each of the edges from a complete graph. These digraphs are called *tournaments*.

Definition 7.3.1 A digraph $T_n = (V, A)$ is a ***tournament*** with n vertices if the underlying graph is the complete graph K_n.

These specific digraphs are called tournaments because they model round-robin-style tournaments in which every team plays every other team exactly once. The direction of an arc can be used to indicate which team won; for example, $a \to b$ if and only if Team A beats Team B.

Example 7.3.1 Five soccer teams have played in a round-robin tournament, with the results as shown below. Model this information as a digraph and determine if a clear winner can be declared.

Team	Teams they Beat
Aardvarks	Bears, Cougars, Eagles
Bears	Cougars, Ducks
Cougars	Ducks, Eagles
Ducks	Aardvarks
Eagles	Bears, Ducks

Solution: The digraph below represents the results from the round-robin tournament. Note that if the directions were removed from the digraph, we would have K_5. If a winner is declared based on the number of teams beaten, then the Aardvarks would be declared the winners of the tournament.

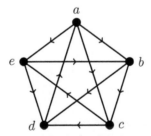

Before we move onto further study of tournaments, one note of caution. In Section 2.3, we used digraphs to model asymmetric relationships for traveling between cities. The digraphs we used were not tournaments since two arcs existed between any pair of vertices (both $a \to b$ and $b \to a$ existed in the digraph). In a tournament, only one of these edges can be present. If a tournament were used to model an instance of the Traveling Salesman Problem, it would represent a scenario in which there is only one direction available for travel between two cities. Although we will still discuss the existence of Hamiltonian cycles and paths within tournaments, it should be noted that these are mainly from an academic perspective and not as a useful model of the Traveling Salesman Problem.

Let us return to Example 7.3.1. Although we gave the complete list of which teams won their games, we could have left out any one of the rows and still obtain the same graph due to the fact that each team must play every other team and exactly one team in a pair can win. Moreover, the number of wins for a team can be seen in the digraph by simply computing the out-degree of a vertex. Recall that the out-degree of a vertex, denoted $\deg^+(x)$, is the number of arcs pointing out of x. Thus in the example above, $\deg^+(a) = 3$ and $\deg^+(d) = 1$.

Suppose we are less concerned with the specific outcomes of a round-robin tournament but rather with the relationships from all possible outcomes. In this case, we do not care if a specific team won three games but how many ways there are for any team to win three games. For example, if all of the results of Example 7.3.1 stayed the same except the Bears beat the Aardvarks and the Ducks beat the Eagles, then a simple relabeling of the graph (shown below) gives the same structure as above.

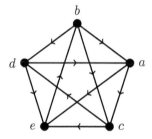

These two graphs are in fact isomorphic (see Section 7.2 for more informa-tion regarding isomorphisms). When comparing the wins for all teams, we see that both graphs had one team that won once, three teams that won twice, and one team that won three times; in short we had wins of $1, 2, 2, 2, 3$. This listing of the wins for a tournament is called a *score sequence*.

Definition 7.3.2 The *score sequence* of a tournament is a listing of the out-degrees of the vertices. It is customary to write these in increasing order.

The score sequence can provide a lot of information about a tournament, but does not give all possible information. For example, consider again the teams from Example 7.3.1. If the Ducks beat the Eagles as opposed to the original outcome of the Eagles beating the Ducks, then we would still have the same score sequence even though the graph itself has changed by one flip of an arc (see below). However, these are quite different outcomes in terms of how you might view a ranking of the teams. In the original scenario (shown below on the left) the three 2-win teams each beat one of the other 2-win teams and the 1-win team. These three teams are in essence interchangeable. However, in the new scenario (shown below on the right), the three 2-win teams have very different win structures and the strength of an opposing team might play into final rankings. For example, the Ducks beat the Aardvarks (3-wins) and the Eagles (1-win), whereas the Cougars beat the Ducks (2-wins) and the Eagles

(1-win). Does this mean the Ducks are a better team than the Cougars since they beat a 3-win team over a 2-win team? As you can see, ranking teams in a round-robin style tournament is non-trivial!

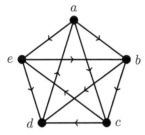 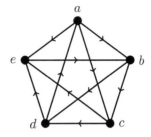

Returning to the notion of a score sequence, it is quite easy to produce such a sequence when the tournament is given to you (simply calculate the out-degrees of the vertices). What is more challenging is producing a tournament given a score sequence. Moreover, how do you know a given sequence could even represent the out-degrees of a tournament?

We begin with some simple necessary conditions for the score sequence of T_n. First note that the maximum out-degree of any vertex is $n-1$ since it can have at most one arc to each of the other vertices in K_n. Moreover, at most one vertex can have out-degree $n-1$ since each pair of vertices has an arc between them. Similarly, the minimum out-degree of any vertex is 0 and at most one vertex can have out-degree 0. Finally, recall that the number of edges in a complete graph K_n is $\frac{n(n-1)}{2}$. In a graph (not digraph) the total degree is always twice the number of edges since each edge adds 2 to the degree count, one for each endpoint. However, as discussed in Theorem 2.10, in a digraph the sum of the out-degrees equals the number of arcs since each arc is only counted once by its direction. Thus in a tournament T_n the sum of the out-degrees equals the number of arcs, which is $\frac{n(n-1)}{2}$. These properties are summarized below.

Properties of Score Sequences
The score sequence of any tournament T_n must satisfy the following:

- $s_1, s_2, \ldots, s_n$ is a sequence of integers satisfying $0 \le s_k \le n-1$ for all $k = 1, 2, \ldots, n$.

- at most one s_k equals 0

- at most one s_k equals $n-1$

- $s_1 + s_2 + \cdots + s_n = \frac{n(n-1)}{2}$

Though these properties are necessary, they are not sufficient; that is, a sequence can satisfy all of the properties listed above yet still not represent the score sequence of a tournament. Below we will discuss two conditions that are both necessary and sufficient and use them to determine if a given sequence is a score sequence of a tournament. The first is closely related to the properties listed above and is easier in its application to a given sequence; the second is more complicated but also provides a method for drawing a tournament with the given score sequence.

Theorem 7.3.3 An increasing sequence $S : s_1, s_2, \ldots, s_n$ (for $n \geq 2$) of nonnegative integers is a score sequence if and only if

$$s_1 + s_2 + \cdots + s_k \geq \frac{k(k-1)}{2}$$

for each k between 1 and n with equality holding at $k = n$.

One additional benefit of this result is that if at any point the inequalities fail to hold, then we do not need to check the remaining inequalities and simply state the sequence is not a score sequence of a tournament.

Example 7.3.2 Determine if the sequence $1, 2, 2, 3, 3, 4$ is the score sequence of a tournament.

Solution: This sequence has length 6, so we will check the inequality above for $k = 1, 2, \ldots 5$, with equality for $k = 6$.

k	$s_1 + \cdots + s_k$	$\frac{k(k-1)}{2}$
1	1	0
2	$1 + 2 = 3$	1
3	$1 + 2 + 2 = 5$	3
4	$1 + 2 + 2 + 3 = 8$	6
5	$1 + 2 + 2 + 3 + 3 = 11$	10
6	$1 + 2 + 2 + 3 + 3 + 4 = 15$	15

Since the inequality $s_1 + \cdots + s_k \geq \frac{k(k-1)}{2}$ holds for each row in the table above, we know that $1, 2, 2, 3, 3, 4$ is a score sequence for T_6.

The next result works by modifying a sequence by removing the last item, which corresponds to deleting one vertex of a tournament and examining the resulting smaller tournament. In theory, this process would continue until either a sequence violates the properties listed above or until a single value remains.

Theorem 7.3.4 An increasing sequence $S : s_1, s_2, \ldots, s_n$ (for $n \geq 2$) of nonnegative integers is a score sequence of a tournament if and only if the sequence $S_1 : s_1, s_2, \ldots, s_{s_n}, s_{s_n+1} - 1, \ldots, s_{n-1} - 1$ is a score sequence.

The new sequence S_1 is created by deleting s_n, rewriting the first s_n terms of S, and then subtracting 1 from any remaining terms. We repeat the process thereby creating shorter sequences. In practice, so long as none of the Score Sequence Properties (see page 239) have been violated, we stop when the sequence reaches length three. There are only two possible tournaments on three vertices, as shown below, and so it is quick to verify if a sequence is a possible score sequence of T_3.

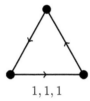

$$1, 1, 1 \qquad\qquad 0, 1, 2$$

Example 7.3.3 Determine if the sequence $1, 2, 3, 3, 3, 3$ is the score sequence of a tournament.

Solution: Let S be the sequence $1, 2, 3, 3, 3, 3$. Then $n = 6$ and $s_n = s_6 = 3$. Thus we form a new sequence S_1 by removing the last term, rewriting the first $s_n = 3$ terms and then subtracting 1 from each of the remaining terms (s_4 and s_5). This produces the sequence S_1 below. Note that we need this sequence in increasing order to continue, so we rewrite this as S_1' below:

$$S_1 : 1, 2, 3, 2, 2 \qquad\qquad S_1' : 1, 2, 2, 2, 3$$

We perform this procedure again on S_1', where now $n = 5$ and $s_5 = 3$. So we get a new sequence S_2 by removing the last term, rewriting the first $s_n = 3$ terms and then subtracting 1 from each of the remaining terms (s_4). This produces the sequence S_2 and its increasing form S_2' below:

$$S_2 : 1, 2, 2, 1 \qquad\qquad S_2' : 1, 1, 2, 2$$

Finally, we perform this procedure one last time on S_2', where $n = 4$ and $s_4 = 2$. We get the sequence

$$S_3 : 1, 1, 1$$

which is one of the two score sequences for T_3. Thus by the theorem above we know that S is the score sequence of a tournament on 6 vertices.

Although this second method is more complex, it has the added benefit of providing a blueprint for how to draw the tournament in question. We will work backwards, beginning by creating the T_3 tournament found, and using the previous score sequences to determine how new vertices and their arcs are added.

Example 7.3.4 Using the results from Example 7.3.3, draw the tournament with score sequence $1, 2, 3, 3, 3, 3$.

Solution: We begin by forming the T_3 tournament with score sequence $1, 1, 1$. To aid in understanding how new vertices are added, we will identify vertices in the digraph with their value in the score sequence at each step.

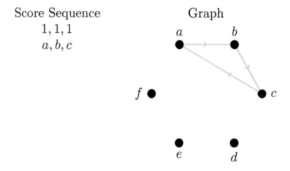

Next we consider the sequence $1, 1, 2, 2$ identified above as S_2'. The addition of d and its arcs is shown below. Note that since the out-degrees of a and b did not change, we know there must be arcs $d \rightarrow a$ and $d \rightarrow b$. Since the out-degree of c increased by 1, we know to add the arc $c \rightarrow d$.

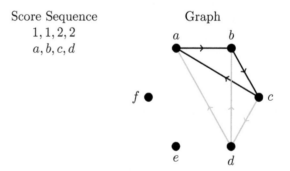

Next we consider the sequence $1, 2, 2, 2, 3$ identified above as S_1'. The addition of e and its arcs is shown below. Note that since the out-degrees of a, c

and d did not change, we know there must be arcs $e \to a, e \to c$ and $e \to d$. Since the out-degree of b increased by 1, we know to add the arc $b \to e$.

Score Sequence   Graph
1, 2, 2, 2, 3
a, b, c, d, e

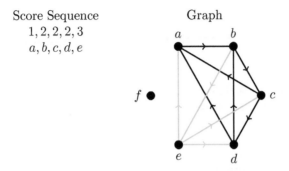

Finally we consider the original sequence $S : 1, 2, 3, 3, 3, 3$. The addition of f and its arcs is shown below. Note that since the out-degrees of a, b and e did not change, we know there must be arcs $f \to a, f \to b$ and $f \to e$. Since the out-degree of c and d increased by 1, we know to add the arcs $c \to f$ and $d \to f$.

Score Sequence   Graph
1, 2, 3, 3, 3, 3
a, b, c, d, e, f

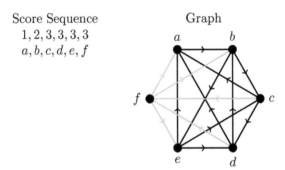

There is one special type of tournament in which no directed cycles exist, referred to as a *transitive* tournament. These tournaments have a very unique score sequence, namely the one consisting of distinct entries.

Definition 7.3.5 A tournament T_n is **transitive** if it does not contain any directed cycles. It has score sequence $0, 1, 2, \ldots, n - 1$.

Transitive tournaments are unique, up to vertex relabeling, and provide an example of a digraph that does not contain a Hamiltonian cycle (see Chapter 2). In fact, if a digraph contains any vertex with out-degree 0, then it cannot possibly contain a Hamiltonian cycle since once that vertex is reached there is no way to exit the vertex. However, transitive tournaments do contain a Hamiltonian path, which can be found by simply following the vertices by decreasing order of out-degree. Surprisingly, the added structure of tournaments over general digraphs is enough to guarantee the existence of a Hamiltonian path in every tournament.

Based on our discussions from Chapter 2, it should come as no surprise that it is hardly trivial to determine if a digraph has a Hamiltonian cycle. As with graphs, there are conditions that can guarantee the existence of a Hamiltonian cycle in a digraph but these are not required properties. One such condition looks at directed paths between vertices in the digraph.

Definition 7.3.6 A digraph D is **strong** if for every pair of vertices x and y in D there is both a directed path from x to y and a directed path from y to x.

It has been shown that if a digraph is strong then a Hamiltonian cycle exists. However, determining if a general digraph is strong can be quite tricky. One way to do this is to check every pair of vertices and search for the appropriate directed paths. Unfortunately, this process is unwieldy since the number of calculations grows very quickly as the number of vertices increases in a graph. Luckily, for tournaments we need only modify one of our previous procedures on score sequences to determine if the tournament is strong, thus allowing us to know if the tournament has a Hamiltonian cycle.

Theorem 7.3.7 An increasing sequence $S : s_1, s_2, \ldots, s_n$ (for $n \geq 2$) of nonnegative integers is a score sequence of a strong tournament if and only if

$$s_1 + s_2 + \cdots + s_k > \frac{k(k-1)}{2}$$

for each k between 1 and $n-1$ with equality holding at $k = n$.

Note that the tournament from Example 7.3.2 satisfied the condition above. Thus it is a strong tournament and therefore contains a Hamiltonian cycle. One possible cycle is shown below in blue.

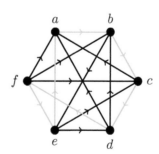

The theorem below summarizes our results for Hamiltonian paths and cycles in tournaments.

Theorem 7.3.8 Let T_n be a tournament on n vertices. Then

- T_n has a Hamiltonian path if $n \geq 2$; and

- T_n has a Hamiltonian cycle if $n \geq 3$ and T_n is strong.

Exercises

7.3.1 There are four possible tournaments on 4 vertices. Draw them and give their score sequences.

7.3.2 Determine if each of the following is a score sequence of a tournament.
(a) $1, 1, 2, 3, 3$
(b) $1, 1, 2, 2, 4$
(c) $0, 2, 2, 2, 3$
(d) $1, 2, 2, 2, 3$
(e) $0, 2, 2, 3, 3, 5$
(f) $1, 1, 2, 3, 4, 4$
(g) $1, 2, 2, 2, 2, 5$
(h) $1, 1, 1, 3, 4, 5$

7.3.3 For each of those from Exercise 2 that is a score sequence, draw the tournament that it represents.

7.3.4 For each of those from Exercise 2 that is a score sequence, determine if the tournament is strong.

7.3.5 For each of those from Exercise 2 that is a score sequence, find a Hamiltonian cycle if it is strong and otherwise find a Hamiltonian path.

7.4 Flow and Capacity

Digraphs have appeared throughout this text to model asymmetric relationships. For example, in Section 2.3 we used digraphs to model changes in cost based on the direction of travel, in Section 3.1 to model one-way streets, and in Section 7.3 to model round-robin tournaments. This section will focus on a new application for digraphs, one in which items are sent through a network. These networks often model physical systems, such as sending water through pipelines or information through a computer network. The digraphs we investigate will most closely resemble those from Section 3.2, as any network will need a starting and ending location, though there is no requirement for the network to be acyclic. In this section, we will need some specialized terminology; in particular, we will use a different notion of a network than the one used in Chapter 4.

Definition 7.4.1 A **network** is a digraph where each arc e has an associated nonnegative integer $c(e)$, called a **capacity**. In addition, the network has a designated starting vertex s called the **source** and a designated ending vertex t called the **sink**. A **flow** f is a function that assigns a value $f(e)$ to each arc of the network.

Below is an example of a network. Each arc is given a two-part label. The first component is the flow along the arc and the second component is the capacity.

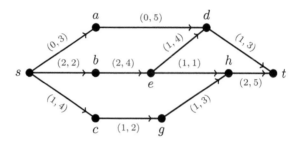

The names of the starting and ending vertices are reminiscent of a system of pipes with water coming from the source, traveling through some configuration of the piping to arrive at the sink (ending vertex). Using this analogy further, we can see that some restraints need to be placed on the flow along an arc. For example, flow should travel in the indicated direction of the arcs, no arc can carry more than its capacity, and the amount entering a junction point (a vertex) should equal the amount leaving. These rules are more formally stated below.

Definition 7.4.2 For a vertex v, let $f^-(v)$ represent the total flow entering v and $f^+(v)$ represent the total flow exiting v. A flow is **feasible** if it satisfies the following conditions:

(1) $f(e) \geq 0$ for all edges e

(2) $c(e) \geq f(e)$ for all edges e

(3) $f^+(v) = f^-(v)$ for all vertices other than s and t

(4) $f^-(s) = f^+(t) = 0$

The notation for in-flow and out-flow mirrors that for in-degree and out-degree of a vertex, though here we are adding the flow value for the arcs entering or exiting a vertex. The requirement that a flow is non-negative indicates the flow must travel in the direction of the arc, as a negative flow would indicate items going in the reverse direction. The final condition listed above requires no in-flow to the source and no out-flow from the sink. This is not necessary in theory, but more logical in practice and simplifies our analysis of flow problems.

The network shown above satisfies the conditions for a feasible flow (check them!). In general, it is fairly easy to verify if a flow is feasible. Our main goal will be to find the best flow possible, called the *maximum flow*.

Definition 7.4.3 The *value* of a flow is defined as $|f| = f^+(s) = f^-(t)$, that is the amount exiting the source which must also equal the flow entering the sink. A *maximum flow* is a feasible flow of largest value.

In practice, we use integer values for the capacity and flow, though this is not required. In fact, given integer capacities there is no need for fractional flows.

Look back at the flow shown in the network above, which has value 3. If we compare the flow and capacity along the arcs, we should see many locations where the flow is below capacity. However, finding a maximum flow is not as simple as putting every arc at capacity — this would likely violate one of the feasibility criteria. For example, if we had a flow of 5 along the arc ad, we would need the flow along dt to also equal 5 to satisfy criteria (3). But in doing so we would violate criteria (2) since the capacity of dt is 3.

The main question in regard to network flow is the optimization question — what is the value of a maximum flow? We could start with a simple feasible flow, as shown above, and use trial and error to keep improving it, though this is not an efficient procedure and does not guarantee the flow we find is indeed maximum. We will discuss an algorithm that not only finds a maximum flow but also provides proof that a larger flow cannot be found. Before we fully discuss the algorithm, we need two additional definitions relating to the flow along a network.

Definition 7.4.4 Let f be a flow along a network. The *slack* k of an arc is the difference between its capacity and flow; that is, $k(e) = c(e) - f(e)$.

Slack will be useful in identifying locations where the flow can be increased. For example, in the network above $k(sa) = 3, k(sc) = 3$ and $k(sb) = 0$ indicates that we may want to increase flow along the arcs sa and sc but no additional flow can be added to sb. The difficult part is determining where to make these additions. To do this we will build special paths, called *chains*, that indicate where flow can be added.

Definition 7.4.5 A *chain* K is a path in a digraph where the direction of the arcs are ignored.

In the network shown above, both $s\,a\,d\,t$ and $s\,a\,d\,e\,h\,t$ are chains, though only $s\,a\,d\,e\,h\,t$ is not a directed path since it uses the arc ed in reverse direction.

We now have all the needed elements for finding the maximum flow. The algorithm below is similar to Dijkstra's Algorithm from Chapter 3 which found the shortest path in a graph (or digraph). Vertices will be assigned two-part labels that aid in the creation of a chain on which the flow can be increased. The format of the Augmenting Flow Algorithm, described below, is an adaption from [39].

Augmenting Flow Algorithm

Input: Network $G = (V, E, c)$, where each arc is given a capacity c, and a designated source s and sink t.

Steps:

1. Label s with $(-, \infty)$

2. Choose a labeled vertex x.

    (a) For any arc yx, if $f(yx) > 0$ and y is unlabeled, then label y with $(x^-, \sigma(y))$ where $\sigma(y) = \min\{\sigma(x), f(yx)\}$.

    (b) For any arc xy, if $k(xy) > 0$ and y is unlabeled, then label y with $(x^+, \sigma(y))$ where $\sigma(y) = \min\{\sigma(x), k(xy)\}$.

3. If t has been labeled, go to Step 4. Otherwise, choose a different labeled vertex that has not been scanned and go to Step 2. If all labeled vertices have been scanned, then f is a maximum flow.

4. Find an $s - t$ chain K of slack edges by backtracking from t to s. Along the edges of K, increase the flow by $\sigma(t)$ units if they are in the forward direction and decrease by $\sigma(t)$ units if they are in the backward direction. Remove all vertex labels except that of s and return to Step 2.

Output: Maximum flow f.

In Step 2 we are labeling the neighbors of a vertex x. We first consider arcs into x from unlabeled vertices that have positive flow (part a) then the arcs out of x to unlabeled vertices with positive slack. These are used to find a chain from s to t onto which flow can be added.

Example 7.4.1 Apply the Augmenting Flow Algorithm to the network on page 245.

Solution:

Step 1: Label s as $(-, \infty)$.

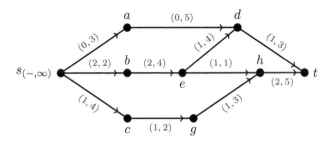

Step 2: Let $x = s$. As there are no arcs to s we will only consider the arcs out of s, of which there are three: sa, sb and sc, which have slack of 3, 0, and 3, respectively. We label a with $(s^+, 3)$, b is left unlabeled since there is no slack on sb, and c is labeled $(s^+, 3)$.

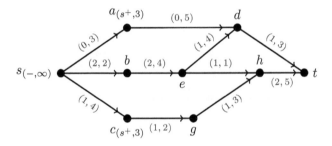

Step 3: As t is not labeled, we will scan either a or c; we choose to start with c. Since the only arc going into c is from a labeled vertex, we need only consider the edges out of c, of which there is only one, cg, with slack of 1. Label g as $(c^+, 1)$ since $\sigma(g) = \min\{\sigma(c), k(cg)\} = \min\{3, 1\} = 1$.

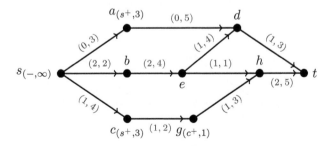

Step 4: As t is not labeled, we will scan either a or g; we choose g. Since the only arc going into g is from a labeled vertex, we need only consider the edges out of g, of which there is only one, gh, with slack of 2. Label h as $(g^+, 1)$ since $\sigma(h) = \min\{\sigma(g), k(gh)\} = \min\{1, 2\} = 1$.

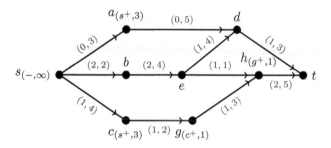

Step 5: As t is not labeled, we will scan either a or h; we choose h. There is one unlabeled vertex with an arc going into h, namely e, which gets a label of $(h^-, 1)$ since $\sigma(e) = \min\{\sigma(h), f(he)\} = \min\{1, 1\} = 1$. The only arc out of h is ht, with slack of 3. Label t as $(h^+, 1)$ since $\sigma(t) = \min\{\sigma(h), k(ht)\} = \min\{1, 3\} = 1$.

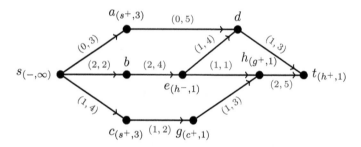

Step 6: Since t is now labeled, we find an $s - t$ chain K of slack edges. Backtracking from t to s gives the chain $scght$, and we increase the flow by $\sigma(t) = 1$ units along each of these edges since all are in the forward direction. We update the network flow and remove all labels except that for s.

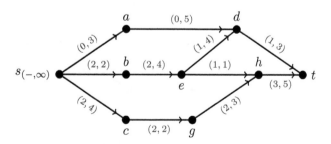

Step 7: As before we will only label the vertices whose arcs from s have slack. We label a with $(s^+, 3)$ and c with $(s^+, 2)$.

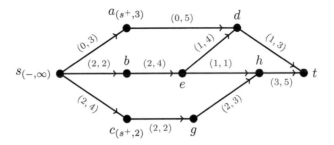

Step 8: We scan either a or c; we begin with c. The only arc from c is to g, but since there is no slack we do not label g. Scanning a we only consider the arc ad, which has slack 3. Label d with $(a^+, 3)$ since $\sigma(d) = \min\{\sigma(a), k(ad)\} = \min\{3, 3\} = 3$.

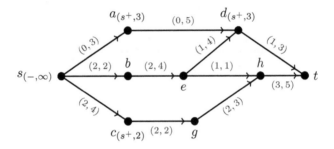

Step 9: Our only unscanned labeled vertex is d. Since e has an arc to d with positive flow, label e as $(d^-, 1)$ since $\sigma(e) = \min\{\sigma(d), f(ed)\} = \min\{3, 1\} = 1$. Also label t with $(d^+, 2)$ since $\sigma(t) = \min\{\sigma(g), k(dt)\} = \min\{3, 2\} = 2$.

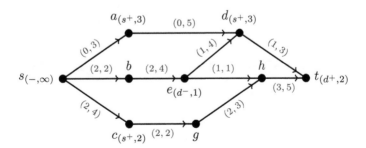

Step 10: Since t is again labeled, we find an $s - t$ chain K' of slack edges. Backtracking from t to s gives the chain $s\,a\,d\,t$, and we increase the flow by $\sigma(t) = 2$ units along each of these edges since all are in the forward direction. We update the network flow and remove all labels except that for s.

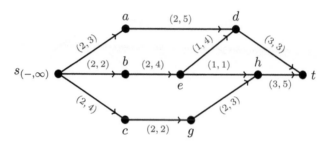

Step 11: The beginning process of again assigning labels is quite similar to the steps above. Some σ values now change along the last chain from which we adjusted the flow. Instead of working through the individual steps, we will pick up after the label for d has been assigned, which is shown below.

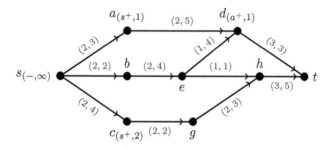

Label e as $(d^-, 1)$ as before since there is flow along the arc ed, but t is not given a label due to no slack along the arc dt. Once this is complete, b will be given a label of $(e^-, 1)$ since $\sigma(b) = \min\{\sigma(e), f(be)\} = \min\{1, 2\} = 1$. No label will be assigned to h since there is no slack along the arc eh. Recall that g also remains unlabeled since cg has no slack.

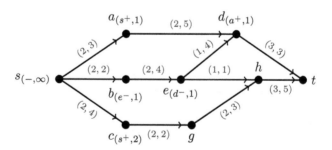

At this point there are no further vertices to label and so f must be a maximum flow, with a value of 6.

When the Augmenting Flow Algorithm halts, a maximum flow has been achieved, though understanding why this flow is indeed maximum requires additional terminology and results. The main idea will be to determine a barrier through which all flow must travel and as a consequence the maximum flow cannot exceed the barrier with minimum size. The source and sink will be on opposite sides of this barrier, which is more commonly called a *cut*.

Definition 7.4.6 Let P be a set of vertices and $\overline{P}$ denote those vertices not in P (called the complement of P). A **cut** $(P, \overline{P})$ is the set of all arcs xy where x is a vertex from P and y is a vertex from $\overline{P}$. An **s − t cut** is a cut in which the source s is in P and the sink t is in $\overline{P}$.

In the network above, if we let $P = \{s, a, e, g\}$ then $\overline{P} = \{b, c, d, h, t\}$ and $(P, \overline{P}) = \{sb, sc, ad, ed, eh, gh\}$. Note that be is not part of the cut even though b and e are in opposite parts of the vertex set (namely b is in $\overline{P}$ and e is in P) since the arc travels in the wrong direction with regards to the definitions of P and $\overline{P}$.

As this cut acts as a barrier to increasing values of a flow, when we discuss the value of a cut we are in fact concerned with the capacities along these arcs rather than their flow. Thus the value of a cut is referred to as its capacity.

Definition 7.4.7 The **capacity** of a cut, $c(P, \overline{P})$, is defined as the sum of the capacities of the arcs that comprise the cut.

The cut given above has capacity 18 (try it!) and therefore indicates that all feasible flows must have value at most 18. Obviously, this is not the best bound for the maximum flow since our work above seems to indicate that the maximum flow has value 6. In fact, two easy cuts often provide more useful initial bounds on the value of a flow; the first is where P only consists of the source and the second is where P consists of every vertex except the sink. In the example above, if we let $P = \{s\}$ then the capacity of this cut is $c(P, \overline{P}) = 9$ and if $P = \{s, a, b, c, d, e, g, h\}$ then it has capacity $c(P, \overline{P}) = 8$, which are much closer to our conjecture that the maximum flow is 6.

The König-Egerváry Theorem from Chapter 5 stated the size of a maximum matching in a bipartite graph equals the size of a minimum vertex cover and allowed us to prove we had a maximum matching by finding a vertex cover of the same size. A similar result holds for flows and cuts in a network.

Theorem 7.4.8 (Max Flow–Min Cut) In any directed network, the value of a maximum $s − t$ flow equals the capacity of a minimum $s − t$ cut.

The difficulty in using this result to prove a flow is maximum is in finding the minimum cut. Luckily, as with the Augmenting Path Algorithm for matching, we can use the vertex labeling procedure to obtain our minimum cut.

Min-Cut Method

1. Let $G = (V, A, c)$ be a network with a designated source s and sink t and each arc is given a capacity c.

2. Apply the Augmenting Flow Algorithm.

3. Define an $s - t$ cut $(P, \overline{P})$ where P is the set of labeled vertices from the final implementation of the algorithm.

4. $(P, \overline{P})$ is a minimum $s - t$ cut for G.

By finding a flow and cut with the same value, we now have proof that the flow is indeed maximum.

Example 7.4.2 Use the Min-Cut Method to find a minimum $s - t$ cut for the network on page 245.

Solution: The final labeling from the implementation of the Augmenting Flow Algorithm produced the network below.

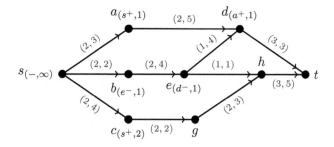

The Min-Cut Method sets $P = \{s, a, b, c, d, e\}$ and $\overline{P} = \{g, h, t\}$. The arcs in the cut are $\{dt, eh, cg\}$, making the capacity of this cut $c(P, \overline{P}) = 3 + 1 + 2 = 6$. Since we have found a flow and cut with the same value, we know the flow is maximum and the cut is minimum.

In practice, we can perform the Augmenting Flow Algorithm and the Min-Cut Method simultaneously, thus finding a maximum flow and providing a proof that it is maximum (through the use of a minimum cut) in one complete procedure.

Exercises

For each of the problems below, use the Augmenting Flow Algorithm to maximize the flow and the Min-Cut Method to find a minimum cut.

7.4.1

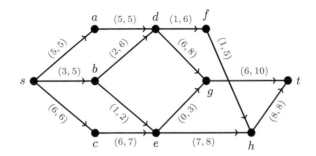

7.4.2

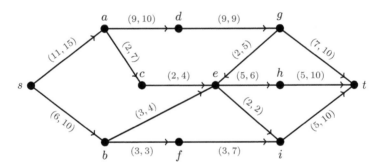

7.4.3

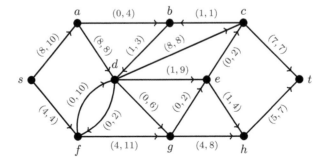

7.4.4

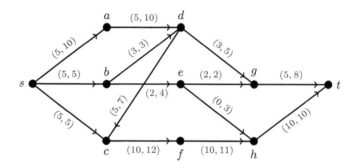

7.4.5

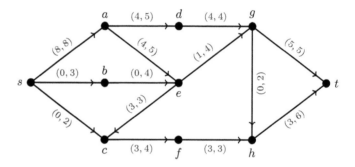

7.5   Rooted Trees

Rooted trees were introduced in Example 4.3 as a method for storing information. This section will further the discussion of, as well as provide additional applications of, rooted trees.

Definition 7.5.1 A **rooted tree** is a tree T with a special designated vertex r, called the **root**. The **level** of any vertex in T is defined as the length of its shortest path to r. The **height** of a rooted tree is the largest level for any vertex in T.

Example 7.5.1 Find the level of each vertex and the height of the rooted tree shown below.

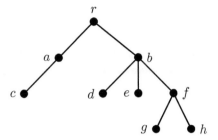

Solution: Vertices a and b are of level 1, c, d, e and f of level 2, and g and h of level 3. The root r has level 0. The height of the tree is 3.

Most people have encountered a specific type of rooted tree: a family tree. The root of a family tree would be the person for whom the descendants are being mapped and the level of a vertex would represent a generation; see the tree below. With this application in mind, the terminology below is used to describe how various vertices are related within a rooted tree.

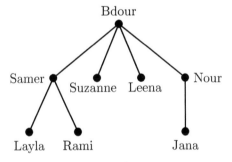

Definition 7.5.2 Let T be a tree with root r. Then for any vertices x and y

- x is a ***descendant*** of y if y is on the unique path from x to r;

- x is a ***child*** of y if x is a descendant of y and exactly one level below y;

- x is an ***ancestor*** of y if x is on the unique path from y to r;

- x is a ***parent*** of y if x is an ancestor of y and exactly one level above y;

- x is a ***sibling*** of y if x and y have the same parent.

Using tree from Example 7.5.1 above, we see that the parent of a is the root r and c is the only child for a. Also, b is the parent of e, but e has no children. The ancestors of g are f, b and r since the unique path from g to r is $g\,f\,b\,r$. The descendants of b are d, e, f, g and h, and the siblings of d are e and f since they all have b as their parent.

Example 4.3 produced a specific type of rooted tree, reproduced below, called a **binary search tree**. The binary portion of the name comes from the fact that each parent in the tree has at most two children. As previously mentioned, binary search trees allow for quick access of information using fewer comparisons than if the information was given in list form. The remainder of this section will focus on two other types of rooted trees that have a strong connection to computer science: breadth-first search trees and depth-first search trees.

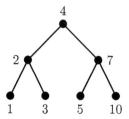

In Chapter 4 we were concerned with finding a minimum spanning tree, often for the use of a specific application needing to connect items at a minimum cost. Here we use search trees to find paths within a graph from a specified root. The applications of these are mainly still within the realm of graph theory, such as finding connected components or bridges within a graph or testing if a graph is planar (see Section 7.6). However, both search trees we will discuss arose, in part, as a way to solve a maze, and have applications into the study of artificial life. Depth-first search trees are credited to the nineteenth century French author and mathematician Charles Pierre Trémaux, and breadth-first search trees were introduced in the 1950s by the American mathematician E.F. Moore.

Depth-First Search Tree

The main idea behind a depth-first tree is to travel along a path as far as possible from the root of a given graph. If this path does not encompass the entire graph, then branches are built off this central path to create a tree. The formal description of this algorithm relies on an ordered listing of the neighbors of each vertex and uses this order when adding new vertices to the tree. For simplicity, we will always use an alphabetical order when considering neighbor lists.

Depth-First Search Tree

Input: Simple graph $G = (V, E)$ and a designated root vertex r.

Steps:

1. Choose the first neighbor x of r in G and add it to $T = (V, E')$.

2. Choose the first neighbor of x and add it to T. Continue in this fashion — picking the first neighbor of the previous vertex to create a path P. If P contains all the vertices of G, then P is the depth-first search tree. Otherwise continue to step (3).

3. Backtrack along P until the first vertex is found that has neighbors not in T. Use this as the root and return to step (1).

Output: Depth-first search tree T.

In creating a depth-first search tree, we begin by building a central spine from which all branches originate. These branches are as far down on this path as possible. In doing so, the resulting rooted tree is often of large height and is more likely to have more vertices at the lower levels.

Example 7.5.2 Find the depth-first search tree for the graph below with the root a.

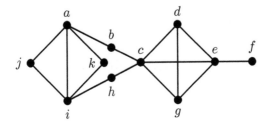

Solution:

Step 1: Since a is the root, we add b as it is the first neighbor of a. Continuing in this manner produces the path shown below. Note this path stops with f since f has no further neighbors in G.

Step 2: Backtracking along the path above, the first vertex with an unchosen neighbor is *e*. This adds the edge *eg* to *T*.

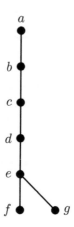

Step 3: Backtracking again along the path from step 1, the next vertex with an unchosen neighbor is *c*. This adds the path *c h i j* to *T*.

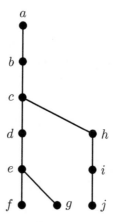

Step 4: Backtracking again along the path from step 3, the next vertex with an unchosen neighbor is *i*. This adds the edge *ik* to *T* and completes the depth-first search tree as all the vertices of *G* are now included in *T*.

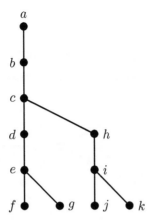

Output: The tree above is the depth-first search tree.

Note that the tree created above has height 5 and with one vertex each at level 1 and 2, two vertices each at level 3 and 4, and four vertices at level 5.

Breadth-First Search Tree

The main objective for a breadth-first search tree is to add as many neighbors of the root as possible in the first step. At each additional step, we are adding all available neighbors of the most recently added vertices.

Breadth-First Search Tree

Input: Simple graph $G = (V, E)$ and a designated root vertex r.

Steps:

1. Add all the neighbors of r in G to $T = (V, E')$.

2. If T contains all the vertices of G, then we are done. Otherwise continue to step (3).

3. Beginning with the first vertex x of r that has neighbors not in T, add all the neighbors of x to T. Repeat this for all the neighbors of r.

4. If T contains all the vertices of G, then we are done. Otherwise repeat step (3) with the vertices just previously added to T.

Output: Breadth-first tree T.

As with depth-first, we will use an alphabetical order when considering neighbor lists. At each stage we are adding a new level to the tree and visually we will place the vertices from left to right, thus aiding in the next stage of vertex additions.

Example 7.5.3 Find the breadth-first search tree for the graph below with the root a.

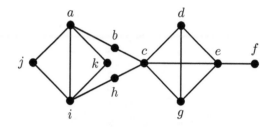

Solution:

Step 1: Since a is the root, we add all of the neighbors of a to T.

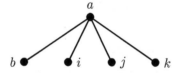

Step 2: We next add all the unchosen neighbors of b, i, j and k, beginning with b as it is the first neighbor of a that was added in Step 1. This adds the edge bc. Moving to i we add the edge ih. No other edges are added since j and k do not have any unchosen neighbors.

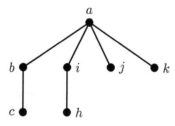

Step 3: We next add all the unchosen neighbors of c, namely d, e and g. No other vertices are added since all the neighbors of h are already part of the tree T.

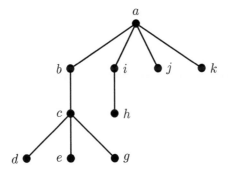

Step 4: Since d has no unchosen neighbors, we move ahead to adding the unchosen neighbors of e. This completes the breadth-first search tree as all the vertices of G are now included in T.

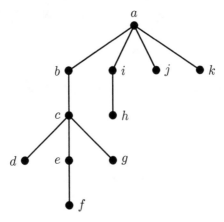

Output: The tree above is the breadth-first search tree.

It should come as no surprise that breadth-first search trees are likely to be of shorter height than their depth-first tree counterpart. The breadth-first search tree created above has height 4 with four vertices on level 1, two vertices on level 2, three vertices on level 3, and one on level 4.

The main difference between these two algorithms is that depth-first focuses on traveling as far into the graph in the beginning, whereas the breadth-first focuses on building outward using neighborhoods. For those who have read Section 7.1, both depth-first and breadth-first algorithms belong in the complexity class P.

Exercises

7.5.1 For the rooted tree T below, with root r, identify the following:
   (a) Level of r, f, h, and k.
   (b) The height of T.
   (c) Parents of r, b, c, f, i.
   (d) Children of r, b, c, f, i.
   (e) Ancestors of r, a, d, h, k.
   (f) Descendants of r, a, d, h, k.
   (g) Siblings of a, f, h, i.

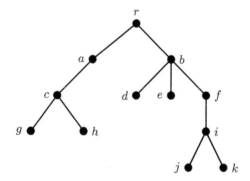

7.5.2 Complete each of the following on the two graphs shown below.
   (a) Find the breadth-first search tree with root a.
   (b) Find the breadth-first search tree with root i.
   (c) Find the depth-first search tree with root a.
   (d) Find the depth-first search tree with root i.

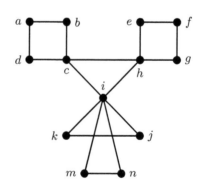

G_1

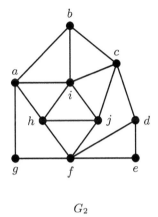

G_2

7.5.3 Explain how to convert a maze into a graph and how breadth-first and depth-first search trees can be used to find a solution.

7.5.4 If the graph is connected, what will the output of depth-first search and breadth-first search be? If the graph is disconnected?

7.6   Planarity

In Section 6.1, we discussed the famous Four Color Theorem and how it relates to graph coloring. One of the important aspects of the result was that the graphs in question are *planar graphs.*

Definition 7.6.1 A graph G is ***planar*** if and only if the vertices can be arranged on the page so that edges do not cross (or touch) at any point other than at a vertex.

Note that for a graph to be planar, it is only required that at least one drawing exists without edge crossings; it is not required that all possible drawings of the graph be without edge crossings. For example, below are two drawings of the graph K_4 (in the language of Section 7.2, these graphs are isomorphic). The drawing on the left is the more standard way of drawing K_4, and contains one edge crossing (ac and bd cross at a location that is not a vertex); the drawing on the right is a planar drawing of K_4 so that no edge crossings exist.

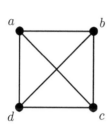
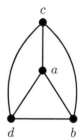

Often it is useful to think of taking a graph and moving around the vertices and pulling or stretching the edges so that they can be repositioned without edge crossings. We could think of obtaining the graph on the right above by rotating the entire graph on the left clockwise by $45°$, moving vertex c above a, and then repositioning the edges from c to the other vertices.

Finding a planar drawing of a graph can be very tricky. In fact, simply determining if a graph is planar or not is hardly trivial. This section will just touch on some of the important aspects of planarity. To begin, we will see an application of planarity (other than our previous interest in graph coloring) and follow with a short discussion on techniques for determining planarity. For a more in-depth study of planarity, see [39] or [40].

Example 7.6.1 Three houses are set to be built along a new city block; across the street lie access points to the three main utilities each house needs (water, electricity, and gas). Is it possible to run the lines and pipes underground without any of them crossing?

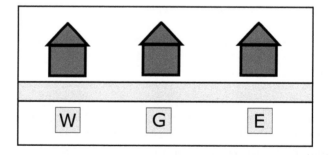

Solution: This scenario can be modeled with three vertices representing the houses (h_1, h_2 and h_3) and three vertices representing the utilities (u_1, u_2 and u_3). First note that if we are not concerned with edge crossings, the proper graph model is $K_{3,3}$, the complete bipartite graph with three vertices in each side of the vertex partition. The standard drawing of $K_{3,3}$, given below, clearly is not a planar drawing as there are many edge crossings; for example, h_1u_2 crosses h_2u_1.

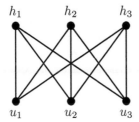

Attempting to find a drawing without edge crossings, we could stretch and move some of the edges as shown below. However, this drawing is still not planar since edges h_3u_1 and h_2u_2 still cross. In fact, no matter how you try to draw $K_{3,3}$ (try it!), there will always be at least one edge crossing. Thus the utility lines and pipes cannot be placed without any of them crossing.

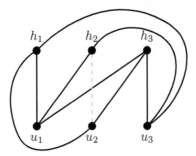

Given a specific drawing of a graph, it is easy to see if that drawing is planar or not. However, just because you cannot find a planar drawing of a graph does not mean a planar drawing does not exist. This is perhaps the most challenging part of planarity. Luckily, we have already seen one of the most important structures in showing a graph is nonplanar (namely $K_{3,3}$). The other structure we studied thoroughly in Chapter 2, namely K_5.

Example 7.6.2 Determine if K_5 is planar. If so, give a planar drawing; if not, explain why not.

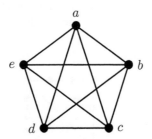

Solution: If we attempt the same procedure as we did for K_4 above, then we run into a problem. We start with the planar drawing of K_4 and add another vertex e below the edge bd (shown on the left below). We can easily add the edges from e to b, c, and d without creating crossings, but to reach a from e we would need to cross one of the edges bd, bc, or cd. Placing vertex e anywhere in the interior of any of the triangular sections of K_4 will still create the need for an edge crossing (one of which is shown on the right below).

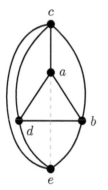
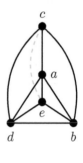

We now have two graphs that we know to be nonplanar. Moreover, having either K_5 or $K_{3,3}$ as a subgraph will guarantee that a graph is nonplanar since if a portion of a graph is nonplanar there is no way for the entire graph to be planar. What may be surprising is that these two graphs provide the basis for determining the planarity of any graph. However, it is not as simple as containing a K_5 or $K_{3,3}$ subgraph, but rather a modified version of these graphs called *subdivisions*.

Definition 7.6.2 A *subdivision* of an edge xy consists of inserting vertices so that the edge xy is replaced by a path from x to y. The subdivision of a graph G is obtained by subdividing edges in G.

Note that a subdivision of a graph can be obtained by subdividing one, two, or even all of its edges. However, the new vertices placed on the edges from G cannot appear in more than one subdivided edge. Below are two examples of subdivided graphs. The graph on the left shows a subdivision of $K_{3,3}$, the graph in the middle shows a subdivision of K_4, the graph on the right shows a subdivision of K_5.

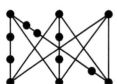

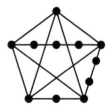

From the discussion above, we should understand why $K_{3,3}$ and K_5 subgraphs pose a problem for planarity. Adding a vertex along any of the edges (thus creating paths between the original vertices) of one of these graphs will not suddenly allow the graph to become planar. Thus containing a subdivision of $K_{3,3}$ or K_5 proves a graph is nonplanar. The Polish mathematician Kazimierz Kuratowski proved in 1930 that containing a $K_{3,3}$ or K_5 subdivision was not only enough to prove a graph was nonplanar, but more surprisingly that any nonplanar graph *must* contain a $K_{3,3}$ or K_5 subdivision.

Theorem 7.6.3 (Kuratowski's Theorem) A graph G is planar if and only if it does not contain a subdivision of $K_{3,3}$ or K_5.

In practice, it is often useful to think of moving vertices and stretching edges at the same time as looking for a $K_{3,3}$ or K_5 subdivision. Note that in order to contain a $K_{3,3}$ subdivision, a graph must have at least 6 vertices of degree 3 or greater, and in order to contain a K_5 subdivision the graph must have at least 5 vertices of degree 4 or greater. These conditions are often helpful when searching for a subdivision.

One final result regarding planarity is quite useful in gaining some intuition as to the planarity of a graph. The following result was proven in 1752 by a mathematician we spent an entire chapter discussing: Leonhard Euler.

Theorem 7.6.4 (Euler's Theorem) If $G = (V, E)$ is a simple planar graph with at least three vertices, then $|E| \leq 3|V| - 6$.

The usefulness of this theorem is not in verifying the relationship between edges and vertices when a graph is known to be planar, but rather in checking if a graph satisfies this inequality. If it does not, then the graph must be nonplanar; however, if the graph satisfies the inequality, then it may or may not be planar.

Example 7.6.3 Determine which of the following graphs are planar. If planar, give a drawing with no edge crossings. If nonplanar, find a $K_{3,3}$ or K_5 subdivision.

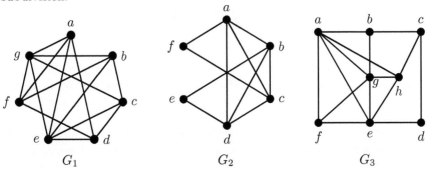

$G_1 \qquad\qquad G_2 \qquad\qquad G_3$

Solution: First note that all of the graphs above are drawn with edge crossings, but this does not indicate their planarity. Also, we can begin by using Euler's Theorem above to give us an indication of how likely the graph is to be planar.

For G_1, we have $|E| = 15$ and $|V| = 7$, giving us the inequality from Euler's Theorem of $15 \leq 3 \cdot 7 - 6 = 15$. Thus the number of edges is as high as possible for a graph with 7 vertices. Although this does not guarantee the graph is nonplanar, it provides good evidence that we should search for a K_5 or $K_{3,3}$ subdivision. Also notice that every vertex has degree at least 4, so we will begin by looking for a K_5 subdivision.

To do this, we start by picking a vertex and looking at its neighbors, hoping to find as many as possible that form a complete subgraph. Beginning with a as a main vertex in K_5, we see the other main vertices would have to be either its neighbors or vertices reachable by a short path. We will start by choosing the other main vertices of the K_5 to be the neighbors of a, namely d, e, f, and g. A starting graph is shown below on the left. Next we fill in the edges between these four vertices, as shown on the right below.

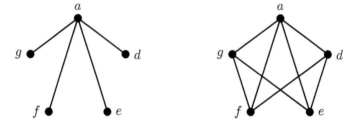

At this point, we are only missing two edges in forming K_5, namely dg and ef. We have two vertices available to use for paths between these unadjacent vertices, and using them we find a K_5 subdivision. Thus G_1 is nonplanar.

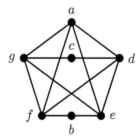

For G_2, we have $|E| = 11$ and $|V| = 6$, giving us the inequality from Euler's Theorem of $11 \leq 3 \cdot 6 - 6 = 12$. We cannot deduce from this result that the graph is nonplanar. However, notice that two vertices have degree 2 and the remaining 4 vertices have degree 4. There cannot be a K_5 subdivision since there are not enough vertices of degree at least 4 and there cannot be a $K_{3,3}$ subdivision since there are not enough vertices of degree at least 3. Thus we can conclude that G_2 is in fact planar. A planar drawing is shown below.

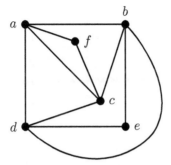

For G_3, we have $|E| = 16$ and $|V| = 8$, giving us the inequality from Euler's Theorem of $16 \leq 3 \cdot 8 - 6 = 18$. As in the previous examples, we cannot deduce that the graph is nonplanar from Euler's Theorem but the high number of edges relative to the number of vertices should give us some suspicion that the graph may not be planar.

When inspecting the vertex degrees, we see only four vertices of degree at least 4 (namely a, e, g, h), indicating the graph cannot contain a K_5 subdivision. However, there are another three vertices of degree 3, allowing for a possibility of a $K_{3,3}$ subdivision.

Again, we start by selecting vertex a to be one of the main vertices of a possible $K_{3,3}$ subdivision. At the same time, we will look for another vertex that is adjacent to three of the neighbors of a. We see that a and g are both adjacent to b, h, e and f. Let us begin with b, h, and e for the vertices on the other side of the $K_{3,3}$, as shown below on the left.

We now search among the remaining vertices (f, d, c) for one that is adjacent to as many of b, h, and e as possible and find c is adjacent to both b and h, as shown below on the right.

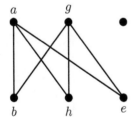
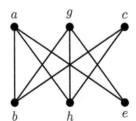

At this point we are only missing one edge to form a $K_{3,3}$ subdivision, namely ce. Luckily we can form a path from c to e using the available vertex d. Thus we have found a $K_{3,3}$ subdivision and proven that G_3 is nonplanar.

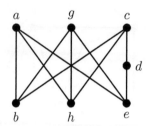

One final note of caution: subdivisions are not necessarily unique. In fact G_1 and G_3 from the example above both contain more than one subdivision, as seen in Exercises 7.6.3 and 7.6.4.

Exercises

7.6.1 Draw a subdivision of K_3 where one edge has two vertices inserted, another edge has one vertex inserted, and the last edge is not subdivided. What is another name for this graph? What can you conclude about C_n, the cycle on n vertices?

7.6.2 Determine if the following graphs are subdivisions of $K_{3,3}$ and explain your answer.

(a)   **(b)**

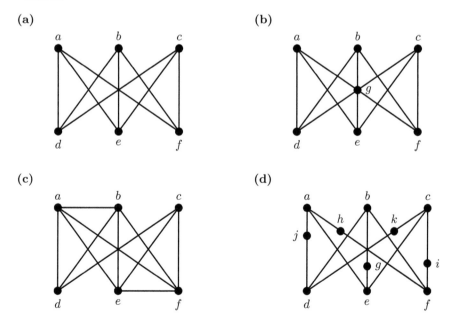

(c)   **(d)**

7.6.3 Using the vertices b, c, e, f, g as the main vertices, find a K_5 subdivision for G_1 from Example 7.6.3.

7.6.4 Find a different $K_{3,3}$ subdivision for G_3 from Example 7.6.3.

7.6.5 For each of the graphs below, determine if it is planar or nonplanar. If planar, give a drawing with no edge crossings. If nonplanar, find a $K_{3,3}$ or K_5 subdivision.

(a) **(b)**

(c) **(d)**

(e)

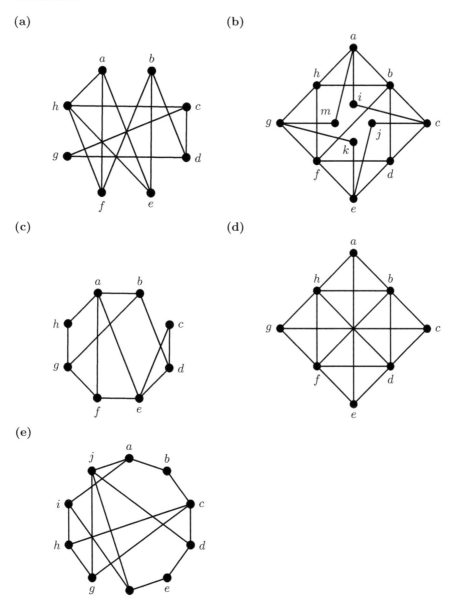

7.7   Edge-Coloring

In Chapter 6 we discussed the applications of graph coloring and investigated the various coloring methods. This section focuses on a different aspect of graph coloring where instead of assigning colors to the vertices of a graph, we will instead assign colors to the edges of a graph. Such colorings are called *edge-colorings* and have their own set of definitions and notations, many of which are analogous to those from Chapter 6. Similar to our previous study of vertex colorings, we will only consider simple graphs, that is graphs without multi-edges. (Note that a graph with a loop cannot be edge-colored).

Definition 7.7.1 Given a graph $G = (V, E)$ an ***edge-coloring*** is an assignment of colors to the edges of G so that if two edges share an endpoint, then they are given different colors. The minimum number of colors needed over all possible edge-colorings is called the ***chromatic index*** and denoted $\chi'(G)$.

Edge colors will be shown throughout this section as various shades of gray and line styles.

Example 7.7.1 Recall that the chromatic number for any complete graph is equal to the number of vertices. Find the chromatic index for K_n for all n up to 6.

Solution: Since K_1 is a single vertex with no edges and K_2 consists of a single edge, we have $\chi'(K_1) = 0$ and $\chi'(K_2) = 1$. Due to their simplicity, a drawing is omitted for these two graphs.

For K_3 since any two edges share an endpoint, we know each edge needs its own color and so $\chi'(K_3) = 3$. For K_4 we can color opposite edges with the same color, thus requiring only 3 colors. Optimal edge-colorings for K_3 and K_4 are shown below.

Edge-coloring K_5 and K_6 is not quite so obvious as those from above. Since no two adjacent edges can be given the same color, we know every edge out of a vertex must be given different colors. Since every vertex in K_5 has degree 4, we know at least 4 colors will be required. Start by using 4 colors out of one of the vertices of the K_5, say a as shown below on the left. Moving to the edges incident to b, we attempt to use our pool of 4 previously used colors; however, one of these is unavailable since it has already been used on the ab edge. A possible coloring of the edges incident to b is given below on the right.

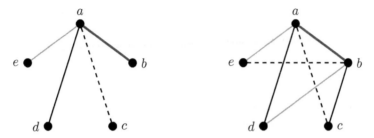

At this point c is left with two incident edges to color but only one color available, namely the one used on edge ab, thus requiring a fifth color to be used. A proper edge-coloring of K_5 is shown below on the left. A similar procedure can be used to find a coloring of K_6 using only 5 colors, as shown below on the right.

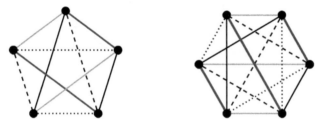

In general, $\chi'(K_n) = n - 1$ when n is even and $\chi'(K_n) = n$ when n is odd.

From the example above we conclude that if a vertex x has degree k, then the entire graph will need at least k colors since each of the edges incident to x will need a different color. Thus the chromatic index must be at least the maximum degree, $\Delta(G)$, of the graph. But notice that for odd values of n, that K_n required one more color than the maximum degree (for example, $\chi'(K_5) = 5$ and $\Delta(K_5) = 4$). In fact, any graph will either require $\Delta(G)$ or $\Delta(G) + 1$ colors to color its edges. This is a much tighter bound than we were able to find for the chromatic number of a graph and was proven in 1964 by the Ukrainian mathematician, Vadim Vizing.

Theorem 7.7.2 (Vizing's Theorem) $\Delta(G) \le \chi'(G) \le \Delta(G) + 1$ for all simple graphs G.

Graphs are referred to as Class 1 if $\chi'(G) = \Delta(G)$ and Class 2 if $\chi'(G) = \Delta(G) + 1$. Some graph types are known to be in each class (for example, bipartite graphs are Class 1 and regular graphs with an odd number of vertices are Class 2). However, with respect to the discussion from Section 7.1, determining which class a graph belongs to is an NP-Complete problem.

For graphs of small size, it is not difficult to find an optimal or nearly optimal edge-coloring. In fact, even using a greedy algorithm (where the first color available is used) will produce an edge-coloring with at most $2\Delta(G) - 1$ colors on any simple graph. More complex algorithms exist that improve on this bound, and if color shifting is allowed then an algorithm exists that will produce an edge-coloring with at most $\Delta(G) + 1$ colors. The example below investigates a suboptimal edge-coloring using a greedy algorithm and explains a better procedure for finding an optimal (or nearly optimal) edge-coloring.

Example 7.7.2 Consider the graph below and color the edges in the order $ac, fg, de, ef, bc, cd, dg, af, bd, bg, bf, ab$ using a greedy algorithm.

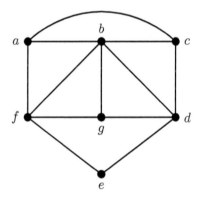

Step 1: Since the first three edges ac, fg, and de are not adjacent, we give each of them the first color.

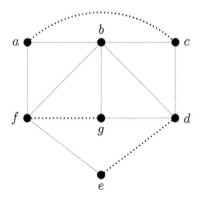

Step 2: Now since ef is adjacent to a previously colored edge, we need a second color for it. Moreover, bc must also use a second color and since these are not adjacent they can have the same color.

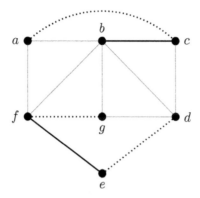

Step 3: Edge cd is adjacent to edges using the first two colors, so a third color is needed. Edge dg can use the second color and edge af must use the third color.

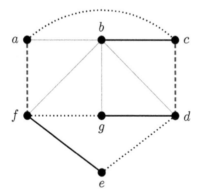

Step 4: Edge bd needs a fourth color since it is adjacent to edges using each of the previous three colors. However, bg can be given the third color.

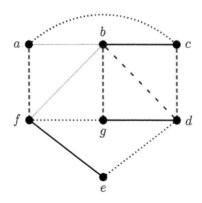

Step 5: Edge bf needs a fifth color since it is adjacent to the first three colors through f and adjacent to the fourth color through b. Moreover, ab needs a sixth color since it is adjacent to the first and third colors through a and the second through fifth colors from b.

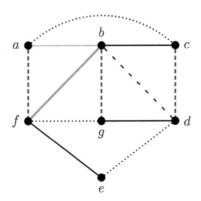

The example above satisfies $\Delta(G) = 5$, and the edge-coloring above uses 6 colors; however $\chi'(G) = 5$. In general, starting with the vertex of highest degree and coloring its edges has a better chance of success in avoiding unnecessary colors, as shown below.

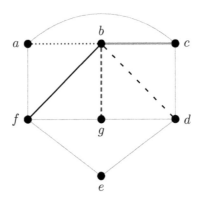

Once we have the minimum number of colors established, we attempt to fill in the remaining edges without introducing an extra color. One possible solution is shown below.

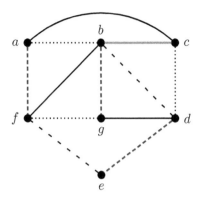

Beyond the analogous definitions and procedures between edge-coloring and vertex-coloring, there is a very direct relationship between the two through the use of a *line graph*.

Definition 7.7.3 Given a graph $G = (V, E)$, the **line graph** $L(G) = (V', E')$ is the graph formed from G where each vertex x' in $L(G)$ represents the edge x' from G and $x'y'$ is an edge of $L(G)$ if the edges x' and y' share an endpoint in G.

Below is the graph from Example 7.7.2 and its line graph. Notice that the vertex e_1 in $L(G)$ is adjacent to e_2 and e_4 through the vertex a in G and e_1 is adjacent to e_3 and e_8 through the vertex c in G.

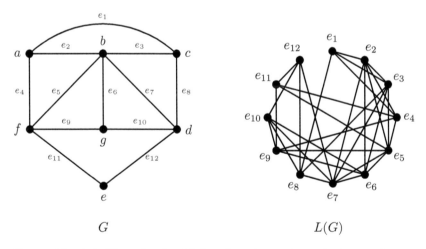

$$G \qquad\qquad\qquad L(G)$$

From this definition, it should be clear how edge-coloring and vertex-coloring are related. In particular, if edge e_1 is given the color blue in G, this would correspond to coloring vertex e_1 in $L(G)$ with blue. This correspondence provides the following result.

Theorem 7.7.4 Given a graph G with line graph $L(G)$, we have $\chi'(G) = \chi(L(G))$.

The result above shows we can find an edge-coloring of any graph by simply vertex-coloring its line graph. However, as we saw in Chapter 6, vertex-coloring is in itself not an easy problem (and in reference to Section 7.1 is in class NP). However, other results on line graphs provide some interest, namely if G is Eulerian then $L(G)$ is Hamiltonian!

Applications of edge-coloring abound, in particular scheduling independent tasks onto machines (different from the scheduling seen in Section 3.2) and communicating data through a fiber-optic network. The example below relates edge-coloring to Section 7.3 and how to schedule games between teams in a round-robin tournament.

Example 7.7.3 The five teams from Example 7.3.1 (Aardvarks, Bears, Cougars, Ducks, and Eagles) need to determine the game schedule for the next year. If each team plays each of the other teams exactly once, determine a schedule where no team plays more than one game on a given weekend.

Solution: Represent each team by a vertex and a game to be played as an edge between the two teams. Then exactly one edge exists between every pair of vertices and so K_5 models the system of games that must be played. Assigning a color to an edge corresponds to assigning a time to a game. By the discussion from Example 7.7.1, we know $\chi'(K_5) = 5$ and so 5 weeks are needed to schedule the games. One such solution is shown below.

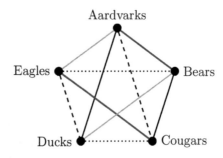

Week	Games	
1	Aardvarks vs. Bears	Cougars vs. Eagles
2	Aardvarks vs. Cougars	Ducks vs. Eagles
3	Aardvarks vs. Ducks	Bears vs. Cougars
4	Aardvarks vs. Eagles	Bears vs. Ducks
5	Bears vs. Eagles	Cougars vs. Ducks

Although the example above is fairly easy to solve as the graph model is a complete graph, the same procedure can be extended to a larger number of teams where each team only plays a subset of all teams in the league. In particular, edge-coloring can be used to determine team schedules in the National Football League!

Ramsey Numbers

As with vertex-coloring, when investigating edge-colorings we are often concerned with finding an optimal coloring (using the least number of colors possible). Other problems exist, where optimality is no longer the goal. One such edge-coloring problem relaxes some of the restrictions on coloring the edges, and is named for the British mathematician and economist, Frank P. Ramsey. Ramsey's legacy is less so due to his own publications, mainly due to his death at the age of 26, but rather for the many theories and results that arose from his limited publications. In particular a minor lemma in his 1928 paper "On a problem of formal logic" stated that within any system there exists some underlying order [32]. Although a simple concept, it birthed an area of mathematics now known as Ramsey Theory.

Ramsey Theory can be described in different forms, so we will naturally use the graph theoretic version. In particular, we will discuss Ramsey numbers as they relate to coloring the edges of a graph. Unlike our edge-colorings above in which no two edges can be given the same color if they have a common endpoint, here we will be concerned with specific monochromatic structures within the larger graph.

Definition 7.7.5 Given positive integers m and n, the **Ramsey number** $R(m, n)$ is the minimum number of vertices r so that all simple graphs on r vertices contain either a clique of size m or an independent set of size n.

Recall that a clique of size m refers to a subgraph with m vertices in which there exists an edge between every pair of vertices. An **independent set** of size n is a group of n vertices in which no edge exists between any two of these n vertices.

To get a better handle on this technical definition, Ramsey numbers are often described in terms of guests at a party. For example, if you wanted to find $R(3, 2)$, then you would be asking how many guests must be invited so that at least 3 people all know each other or at least 2 people do not know each other. Try it!

Ramsey numbers can be viewed as coloring the edges of a complete graph using two colors, say blue and black, so that either a blue clique of size m exists or a black clique of size n exists. You can view the black clique as the edges that would not exist in the graph, thus making their endpoints an independent set of vertices. Proving $R(m, n) = r$ requires two steps: first, we find an edge-coloring of K_{r-1} without a blue m-clique and without a black

n-clique; second, we must show that any edge-coloring of K_r will have either a blue m-clique or a black n-clique. These steps should feel familiar, as they mirror our discussion of the chromatic number in Chapter 6.

Example 7.7.4 Determine $R(3,3)$.

Solution: First note that we are searching for a 3-clique of either color, also known as a monochromatic triangle. However, $R(3,3) > 5$ as the edge-coloring of K_5 below does not contain a monochromatic triangle.

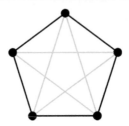

Next, consider K_6. Since each vertex has degree 5 and we are coloring the edges using only two colors, we know every vertex must have at least 3 adjacent edges of the same color. Suppose a has three adjacent blue edges, as shown below. All other edges are shown in gray to indicate we do not care (yet) which color they have.

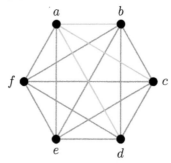

If any one of the edges between b, c and d is blue, then a blue triangle exists using the edges back to a. One possibility is shown below.

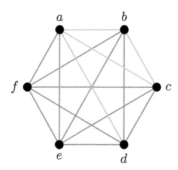

Otherwise, none of the edges between b, c and d are blue, creating a black triangle among these vertices.

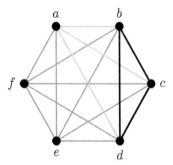

Although the argument above used vertex a with three adjacent blue edges, a similar argument would hold for any vertex of K_6 and for a vertex with three black adjacent edges. Thus $R(3, 3) = 6$.

Using the discussion at the end of the previous example for inspiration, we note two Ramsey number relationships:

- $R(m, n) = R(n, m)$

- $R(2, n) = n$

The first shows that we can interchange m and n without impacting the Ramsey number, as can be seen by simply switching the color on every edge from blue to black and vice versa. The explanation for the second relationship is left as an exercise.

Although the solution for $R(3, 3)$ was not too difficult to determine, increasing m and n both by one value greatly increases the complexity of a solution. In fact, $R(4, 4) = 18$ was proven in 1955 and yet $R(5, 5)$ is still unknown. The table below lists some known values and bounds for small values of m and n at the time of publication. A single value in a column indicates the exact value is known; otherwise, the two values given are the upper and lower bounds as stated. Further discussion and results can be found at [34].

m	3	4	5		6		7	
n			lower	upper	lower	upper	lower	upper
3	6	9	14		18		23	
4		18	25		36	41	49	61
5			43	48	58	87	80	143
6					102	165	115	298
7							205	540

Exercises

7.7.1 Find an optimal edge-coloring for each of the graphs below.

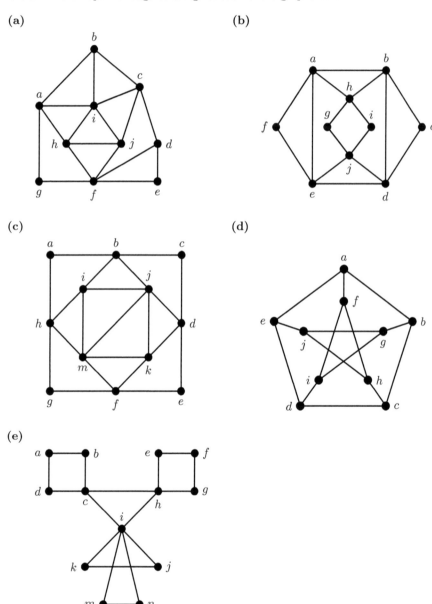

7.7.2 Draw the line graph for each of the graphs above.

7.7.3 Explain why $R(2, n) = n$.

Appendix

In Chapter 4, the software package GeoGebra was mentioned as a tool for finding Fermat points and Steiner trees. This section provides additional instruction on using the program and activities for gaining better understanding of the program. It is broken into three parts that are useful for finding a Fermat point.

When you open GeoGebra, a blank file is started. The ribbon at the top contains tools that either create items (such as points, lines, and polygons) or calculates properties (such as angle measure or length) of previously created items. The Undo button (top right-hand corner of your screen) will be helpful if you select items in the wrong order.

Creating a Triangle

1. Click on the tool that looks like a shaded in triangle. This is the **Polygon** tool.

2. A drop down menu will appear. Choose **Polygon**.

3. Click anywhere on the screen. This plots point A.

4. Click another location for point B.

5. Click a third location for point C.

6. Click on A again to create triangle $\triangle ABC$.

Finding Angle Measure

1. Click on the tool that looked like an angle. This is the **Measure** tool.

2. A drop down menu will appear. Choose **Angle**.
   Note: You can either choose two line segments or three points. It is usually easier to work with the line segments.

3. Click on two edges that share a vertex.
   If you think of the edges as hands on a clock, you must select them in counter-clockwise order to obtain the smaller angle. If you pick them in clockwise order, you can either subtract the answer from 360 or use the undo button and reselect the edges.

4. The angle will appear in green.

Finding the Fermat Point

1. Select the **Polygon** tool.

2. In the drop down menu, select **Regular Polygon**.

3. Select two of the vertices from triangle $\triangle ABC$.
   Unlike with angle measure, you must select the vertices in clockwise order for the new point of the equilateral triangle to be outside $\triangle ABC$. Hit undo and retry if the equilateral triangle is on the wrong side of your chosen edge.

4. A box will pop up asking for number of vertices. Type 3 and hit Enter. A third point and the equilateral point will appear.

5. Repeat the above steps until all three triangles are formed.

6. Select the tool that looks like a line through two points. This is the **Line** tool.

7. In the drop down menu, select **Segment**.

8. Select the new vertex (say X) of an equilateral triangle and the opposite vertex from the original triangle (say C).

9. Repeat Step 8 until all three line segments formed.

10. Select the tool that is a dot with the letter A above it. This is the **Point** tool.

11. In the drop down menu, select **Intersect**.

12. Click on two of the lines or hover over the intersection point until the three segments appear darker and click. This is the Fermat Point.

Other Items

- When the **Move** tool (looks like an arrow) is selected, the box in the far right upper corner (has three lines with a circle and triangle over top) opens up a bar with options for the background. I find having grid lines and axes helpful.

- Right clicking on an object opens up options for how it is viewed.

- Distance of line segments can be useful. Click on Object Properties. A panel will open to the right and the drop down box by Show Label allows you to choose Value. This will label a line segment with its length. Names of vertices can be useful. Click on Show Label.

Exercises

A.1 Create a triangle ABC where $A = (2, 2)$, $B = (8, 4)$ and $C = (6, 6)$.

1. Find the Fermat point P and list its coordinates.

2. Find the distance of each segment from P to the vertices of the triangle and give the total distance.

3. Find the minimum spanning tree and compare it the shortest network for the triangle.

A.2 Plot the following points $A = (1, 1)$, $B = (3, 5)$, $C = (7, 5)$, $D = (8, 6)$, $E = (12, 6)$, $F = (12, 4)$, $G = (14, 8)$.

1. The MST for these points consists of the edges AB, BC, CD, DE, EF, and EG. Add these line segments. Calculate the total weight of the MST.

2. Use the Angle Measure tool to determine locations for improving the MST. Use the procedure above to find the shortcut points. List their coordinates.

3. Calculate the total weight of the new network. Compare it to the total weight of the MST.

Selected Answers and Solutions

Chapter 1: Eulerian Tours

1.2 Euler circuits: (a), (b), (d).

1.3 A graph is Eulerian if and only if it is connected and all vertices are even. A graph is semi-Eulerian if and only if it is connected and exactly two vertices are odd. A graph cannot have all even vertices while still having two odd vertices. Thus a graph cannot be both Eulerian and semi-Eulerian.

1.4 **(a)** (i) $\deg(a) = 4$, $\deg(b) = 4$, $\deg(c) = 4$, $\deg(d) = 2$, $\deg(e) = 6$, $\deg(f) = 3$, $\deg(g) = 4$, $\deg(h) = 3$; (ii) semi-Eulerian
(b) (i) $\deg(a) = 4$, $\deg(b) = 4$, $\deg(c) = 3$, $\deg(d) = 4$, $\deg(e) = 4$, $\deg(f) = 3$, $\deg(g) = 6$, $\deg(h) = 4$, $\deg(i) = 6$, $\deg(j) = 4$; (ii) semi-Eulerian
(c) (i) $\deg(a) = 2$, $\deg(b) = 2$, $\deg(c) = 4$, $\deg(d) = 2$, $\deg(e) = 2$, $\deg(f) = 3$, $\deg(g) = 2$, $\deg(h) = 3$; (ii) neither
(d) (i) $\deg(a) = 3$, $\deg(b) = 5$, $\deg(c) = 3$, $\deg(d) = 5$, $\deg(e) = 3$, $\deg(f) = 5$, $\deg(g) = 3$, $\deg(h) = 5$; (ii) neither
(e) (i) $\deg(a) = 4$, $\deg(b) = 4$, $\deg(c) = 2$, $\deg(d) = 2$, $\deg(e) = 4$, $\deg(f) = 2$; (ii) Eulerian
(f) (i) $\deg(a) = 2$, $\deg(b) = 4$, $\deg(c) = 2$, $\deg(d) = 4$, $\deg(e) = 4$, $\deg(f) = 2$, $\deg(g) = 2$, $\deg(h) = 4$, $\deg(i) = 2$, $\deg(j) = 4$; (ii) Eulerian

1.16 In Step 1, find a path from one odd vertex to the other. Continue finding circuits as described in Step 2, combining them with the initial path to create a trail. Repeat the process until all edges are included in the trail.

1.17 Since the number of edges is 15, the total degree must be 30. The 6 vertices give a total degree of 24 and so the last two vertices have a degree total of 6. Since there must be an odd number of odd vertices, either both or neither of the remaining vertices must have odd degree. The possible degrees are 1 and 5, 2 and 4, and 3 and 3. It cannot be 0 and 6 since the graph is connected.

Chapter 2: Hamiltonian Cycles

2.1 (a) $40,320$
  (b) $479,001,600$
  (c) $20,922,789,888,000$

2.2 (a) $9! = 362,880$
  (b) $11 * 10 * 9 = 990$
  (c) $7! * 6 = 30,240$

2.3 $K_4 :$ $(4-1)! = 3! = 6$
$K_8 :$ $(8-1)! = 7! = 5040;$
$K_{10} :$ $9! = 362,880$

2.5

2.6 (a) (iii) G has 6 vertices, all of which have degree at least 3. If G is connected, then it satisfies Dirac's Theorem and has a Hamiltonian cycle. If G is not connected, then it does not have a Hamiltonian cycle.
  (b) (i) Use Dirac's Theorem.
  (c) (ii) G cannot have a Hamiltonian cycle if it contains a vertex of degree 1.
  (d) (iii) It is possible though not guaranteed.
  (e) (ii) G is not connected.

2.7 *abcdea* - 1290; *badceb* → *adceba* - 1250; *cbadec* → *adecba* - 1190; *dabced* → *abceda* - 1190; *eabcde* → *abcdea* - 1290

2.8 (a) (i) *adbcea* - 21; *bdaceb* → *acebda* - 19; *cedabc* → *abceda* - 18; *dabced* → *abceda* - 18; *ecbade* → *abceda* - 18; (ii) *adbcea* - 21; (iii) *abceda* - 18
  (b) (i) *mqonpm* - 1285; *nmqopn* - 1415; *onmqpo* - 1355; *pmqonp* - 1285; *qmnopq* - 1355; (ii) *mnopqm* - 1355; (iii) *mpnoqm* - 1285
  (c) (i) *acbfdea* - 23; *bcdeafb* - 27; *cbfdeac* - 23; *debcafd* - 27; *edcbfae* - 27; *fbcdeaf* - 27 (answers may vary in the case of ties); (ii) *aedcbfa* - 27; (iii) *acbfdea* - 23
  (d) (i) *fhgkif* - 138.75; *ghikfg* - 152.75; *hgkifh* - 138.75; *ihgkfi* - 135.75; *kghifk* - 135.75; (ii) *fihgkf* - 135.75; (iii) *fhgkif* - 138.75
  (e) (i) *jnkmopj* - 1548; *kmjnpok* - 1442; *mknjopm* - 1483; *njmkopn* - 1442; *opnjmko* - 1351; *pomknjp* - 1548 (ii) *jmkopnj* - 1442; (iii) *jnkmpoj* - 1483

Chapter 3: Paths

3.1 (a) path: $xaby$ weight: 12
(b) path: $xejy$ weight: 6
(c) path: $xbey$ weight: 9
(d) path: $xaegy$ weight: 15
(e) path: $xcdy$ weight: 10
(f) path: $xcjhy$ weight: 10
(g) path: $xhgfemyy$ weight: 12

3.4 (a) path: $xabdgey$ weight: 20
(b) path: $xfjidy$ weight: 21
(c) path: $xbcey$ weight: 14

3.5 The critical path priority list gave a schedule with a finishing time equal to the critical time of *Start*. No additional processors can reduce the finishing time.

Chapter 4: Trees and Networks

4.1 (d) and (e) are both trees; (a), (f), and (g) are all not trees; (b) and (c) may or may not be trees.

4.3 Weights of minimum spanning trees: (a) 11 (b) 65 (c) 21 (d) 28 (e) 1018.

4.5 Kruskal's Algorithm will produce a spanning forest of the disconnected graph; Prim's Algorithm would produce a spanning tree of the component containing the root vertex.

4.6 Insert a step 0 that chooses the required edge. Continue with the algorithms as before. The resulting tree might be minimum but cannot be guaranteed.

4.9 If a graph is very sparse (the number of edges is close to $n-1$), then Reverse Delete would only need a few steps to obtain a tree. However, if a graph is very dense (so the number of edges is much greater than $n-1$), then using Reverse Delete would require many more steps than Kruskal's to obtain a minimum spanning tree.

4.12 Total weight is 1608.

Chapter 5: Matching

5.6 **(a)** (a), (b), (d), and (f) are bipartite.

5.8 One possible solution: Benefits–Agatha, Computing–Leah, Purchasing–George, Recruitment–Dinah, Refreshments–Nancy, Social Media–Evan, Travel Expenses–Vlad.

5.12 **(a)** Rich–Alice, Stefan–Dahlia, Tom–Beth, Victor–Cindy
(b) Alice–Rich, Beth–Stefan, Cindy–Victor, Dahlia–Tom

Chapter 6: Graph Coloring

6.2 (a) 4 (b) 4 (c) 4 (d) 3 (e) 3 (f) 3

6.9 Answers may vary. Optimal number of colors is listed.
(a) 8 colors. $a = \{5,6\}\, b = \{1,2,3\}\, c = \{7,8\}\, d = \{5,6\}\, e = \{1,2,3,4\}\, f = \{7,8\}\, g = \{1,2,3\}\, h = \{4\}$
(b) 10 colors. $a = \{3,4,5,6\}\, b = \{8,9\}\, c = \{3,4,5,6,7\}\, d = \{8,9,10\}\, e = \{1,2\}\, f = \{3,4,5\}\, g = \{1\}\, h = \{7,8,9\}\, i = \{1,2\}\, j = \{10\}$
(c) 11 colors. $a = \{8\}\, b = \{1,2,3,4\}\, c = \{9,10,11\}\, d = \{3,4\}\, e = \{5,6,7\}\, f = \{5,6,7,8\}\, g = \{1,2\}$
(d) 8 colors. $a = \{1,2,3\}\, b = \{4,5,6\}\, c = \{7,8\}\, d = \{1,2\}\, e = \{4,5,6,7\}\, f = \{4,5,6\}\, g = \{1,2\}\, h = \{1,2,3\}\, i = \{3\}, j = \{8\}$
(e) 13 colors. $a = \{10,11,12\}\, b = \{1,2,3\}\, c = \{4,5,6,7\}\, d = \{1,2\}\, e = \{4,5,6,7,8\}\, f = \{9,10,11\}\, g = \{1,2,3,4\}\, h = \{6,7,8,9\}\, i = \{13\}, j = \{9,10,11,12\}\, k = \{6,7,8\}\, m = \{1,2,3,4,5\}$
(f) 11 colors. $a = \{9,10,11\}\, b = \{1,2,3,4,5\}\, c = \{6,7,8\}\, d = \{9,10\}\, e = \{1,2,3,4\}\, f = \{5,6\}\, g = \{1,2\}\, h = \{6,7,8\}\, i = \{1,2,3\}\, j = \{5,6,7\}$

Chapter 7: Additional Topics

7.1.1 Answers may vary
(a) $a = 1, c = 10$
(b) $a = 1, c = 4$
(c) $a = 1, c = 2$
(d) $a = 1, c = 3$
(e) $a = 4, c = 2$

(f) $a = 7, c = 1$

7.2.1 Isomorphic $(a \to s, b \to y, c \to z, d \to w, e \to u, f \to t, g \to x, h \to v)$.

7.2.2 Not isomorphic (look at vertex degrees).

7.2.3 Isomorphic $(a \to x, b \to v, c \to t, d \to w, e \to y, f \to u, g \to z)$.

7.2.4 Isomorphic $(a \to s, b \to v, c \to t, d \to y, e \to u, f \to x, g \to w, h \to z)$.

7.3.1 $0, 1, 2, 3;\ 0, 2, 2, 2;\ 1, 1, 2, 2;\ 1, 1, 1, 3.$

7.3.2 (a), (b), (d), (e), (f) and (h) are tournaments.

7.3.4 (a), (d) and (f) are strong; (b), (e) and (h) are not strong.

7.4.1 Flow $= 16$, $P = \{s\}, \overline{P} = \{a, b, c, d, e, f, g, h, t\}$.

7.4.2 Flow $= 20$, $P = \{s, a, b, c, d\}, \overline{P} = \{e, f, g, h, i, t\}$.

7.4.3 Flow $= 14$, $P = \{s\}, \overline{P} = \{a, b, c, d, e, f, g, h, t\}$.

7.4.4 Flow $= 17$, $P = \{s, a, b, c, d, e, f\}, \overline{P} = \{g, t\}$.

7.4.5 Flow $= 10$, $P = \{s, a, b, c, d, e, f, g\}, \overline{P} = \{h, t\}$.

7.5.1   **(a)** 0; 2; 3; 4
  (b) 4
  (c) none; r; a; b; f
  (d) a, b; d, e, f; g, h; i; j, k
  (e) none; r; r, b; r, a, c; r, b, f, i
  (f) $a, b, c, d, e, f, g, h, i, j, k$; c, g, h; none; none; none
  (g) b; d, e; g; none

7.6.1 C_6; Every C_n (for $n \geq 3$) is a subdivision of K_3.

7.6.2 **(a)** and **(d)** are subdivisions; **(b)** is not since vertex g is on more than one path between vertices; **(c)** is not since extra edges were added to the $K_{3,3}$ graph.

7.6.5 **(a)** nonplanar; **(b)** planar; **(c)** planar; **(d)** planar; **(e)** nonplanar.

7.7.1 (b) 4 (e) 6.

Bibliography

[1] Alexandru T. Balaban. "Applications of Graph Theory in Chemistry." In: *J. Chem. Inf. Comput. Sci.* 25 (1985), pp. 334–343.

[2] Norman L. Biggs, E. Keith Lloyd, and Robin J. Wilson. *Graph Theory 1736-1936*. Oxford: Clarendon Press, 1976. ISBN: 0-19-853901-0.

[3] Eric Bonabeau and Théraulaz. "Swarm Smarts." In: *Scientific American* (Mar. 2000).

[4] J.A. Bondy and U.S.R. Murty. *Graph Theory*. New York: Springer, 2008. ISBN: 978-1-84628-969-9.

[5] Gary Chartrand. *Introductory Graph Theory*. New York: Dover, 1984. ISBN: 978-0486247755.

[6] William Cook. *The Traveling Salesman Problem*. 2015. URL: `http://www.math.uwaterloo.ca/tsp/index.html` (visited on 06/08/2015).

[7] William J. Cook. *In Pursuit of the Traveling Salesman*. Princeton, NJ: Princeton University Press, 2012. ISBN: 978-0-691-15270-7.

[8] G. Dantzig, R. Fulkerson, and S. Johnson. "Solution of a large-scale traveling-salesman problem." In: *J. Operations Res. Soc. Amer.* 2 (1954), pp. 393–410. ISSN: 0160-5682.

[9] Reinhard Diestel. *Graph Theory*. 3rd ed. New York: Springer, 2005. ISBN: 978-3-540-26182-7.

[10] Edsger W. Dijkstra. "A note on two problems in connexion with graphs." In: *Numerishe Mathematik* 1 (1959), pp. 269–271.

[11] Edsger. W. Dijkstra. "Reflections on [[10]]." Circulated privately. 1982. URL: `http://www.cs.utexas.edu/users/EWD/ewd08xx/EWD841a.PDF`.

[12] G.A. Dirac. "Some theorems on abstract graphs." In: *Proc. Lond. Math. Soc.* 2 (1952), pp. 69–81.

[13] Jack Edmonds and Ellis L. Johnson. "Matching, Euler tours and the Chinese postman." In: *Mathematical Programming* 5 (1973), pp. 88–124.

[14] Leonhard Euler. "Solutio problematis ad geometriam situs pertinentis." In: *Commentarii Academiae Scientiarum Imperialis Petropolitanae* 8 (1736), pp. 128–140.

[15] Mark Fernandez. *High Performance Computing*. 2011. URL: http://en.community.dell.com/techcenter/high-performance-computing/w/wiki/2329 (visited on 06/26/2015).

[16] Fleury. "Deux problèmes de géométrie de situation." In: *Journal de mathématiques élémentaires* 2 (1883), pp. 257–261.

[17] D. Gale and L.S. Shapley. "College Admissions and the Stability of Marriage." In: *The American Mathematical Monthly* 69 (1962), pp. 9–15.

[18] Martin Charles Golumbic and Ann N. Trenk. *Tolerance Graphs*. New Tork, NY: Cambridge University Press, 2004. ISBN: 0-521-82758-2.

[19] R.L. Graham and Pavol Hell. "On the history of the minimum spanning tree problem." In: *Annals of the History of Computing* 7 (1985), pp. 43–57.

[20] Dan Gusfield and Robert W. Irving. *The Stable Marriage Problem*. Cambridge, MA: The MIT Press, 2012. ISBN: 0-262-07118-5.

[21] John M. Harris, Jeffry L. Hirst, and Michael J. Mossinghoff. *Combinatorics and Graph Theory*. 2nd ed. New York: Springer, 2008. ISBN: 978-0-387-797710-6.

[22] Carl Hierholzer. "Über die möglichekeit, einen linienzug ohne wiederholung und ohne unterbrechung zu umfahren." In: *Mathematische Annalen* 6 (1873), pp. 30–32.

[23] Intel. *Processors - Intel Microprocessor Export Compliance Marks*. 2014. URL: http://www.intel.com/support/processors/sb/CS-032813.htm (visited on 07/10/2015).

[24] Dieter Jungnickel. *Graphs, Networks and Algorithms*. 4th ed. New York: Springer, 2013. ISBN: 978-3-642-32277-8.

[25] H.A. Kierstead. "A polynomial time approximation algorithm for dynamic storage allocation." In: *Discrete Appl. Math.* 88 (1991), pp. 231–237.

[26] H.A. Kierstead and Karin R. Saoub. "First-Fit coloring of bounded tolerance graphs." In: *Discrete Appl. Math.* 159 (2011), pp. 605–611.

[27] H.A. Kierstead and Karin R. Saoub. "Generalized Dynamic Storage Allocation." In: *Discrete Mathematics & Theoretical Computer Science* 16 (2014), pp. 253–262.

[28] H.A. Kierstead, D. Smith, and W. Trotter. "First-Fit coloring of interval graphs has performance ratio at least 5." In: *European J. of Combin.* 51 (2016), pp. 236–254.

[29] H.A. Kierstead and W. Trotter. "An extremal problem in recursive combinatorics." In: *Congr. Numer.* 33 (1981), pp. 143–153.

[30] Harold Kuhn. "The Hungarian Method for the assignment problem." In: *Naval Research Logistics Quarterly* 2 (1955), pp. 83–97.

[31] Roger Mallion. "A contemporary Eulerian walk over the bridges of Kaliningrad." In: *BSHM Bulletin* 23 (2008), pp. 24–36.

[32] F.P. Ramsey. "On a problem in formal logic." In: *Proc. London Math. Soc.* 30 (1929), pp. 264–286.

[33] Michael Sipser. *Introduction to the Theory of Computation.* 3rd ed. Boston, MA: Cengage Learning, 2013. ISBN: 978-1-133-18779-0.

[34] *Small Ramsey Numbers.* 2017. URL: http : / / www . combinatorics . org / ojs / index . php / eljc / article / view / DS1 / pdf (visited on 03/11/2017).

[35] Peter Tannenbaum. *Excursions in Modern Mathematics.* 8th ed. Boston, MA: Pearson, 2014. ISBN: 978-0-3218-2573-5.

[36] *The Merriam Webster Dictionary.* Springfield, MA: Merriam Webster, 2005. ISBN: 978-0-8777-9636-7.

[37] *The Sveriges Riksbank Prize in Economic Science in Memory of Alfred Nobel.* 2013. URL: http://www.nrmp.org/wp-content/uploads/2013/ 08/The-Sveriges-Riksbank-Prize-in-Economic-Sciences-in- Memory-of-Alfred-Nobel1.pdf (visited on 11/05/2015).

[38] Top500.org. *High Performance Computing.* 2014. URL: http://top500. org/ (visited on 06/26/2015).

[39] Alan Tucker. *Applied Combinatorics.* 6th ed. Hoboken, NJ: Wiley, 2012. ISBN: 978-0-470-45838-9.

[40] Douglas B. West. *Introduction to Graph Theory.* 2nd ed. Upper Saddle River, NJ: Prentice Hall, 2001. ISBN: 0-13-014400-2.

[41] Robin Wilson. *Four Colors Suffice.* Princeton, NJ: Princeton University Press, 2002. ISBN: 0-691-11533-8.

Image Credits

Most of the figures that appear in this book were created electronically by the author. Special thanks are due to David Taylor for his assistance in creating the graphs and tables throughout this book.

For the other images that were used, either with explicit permission or via public domain use, credit is given here, organized by order of appearance.

- The map of Königsberg on page 1 is a public domain image, file *Image-Koenigsberg,_Map_by_Merian-Erben_1652.jpg*.

- The drawing of the bridges of Königsberg on page 4 appeared in [14] and is a public domain image.

- The drawing of bridges within a city on page 32 appeared in [14] and is a public domain image.

- The map on page 209 is a public domain image, courtesy of author Jkan997 of Wikimedia Commons, file *Amtrak_Cascades.svg*, released under the Creative Commons Attribution-ShareAlike 3.0 Unreported License.

Index